Industrial Electrical Troubleshooting

Industrial Electrical Troubleshooting

LYNN LUNDQUIST

DELMAR
CENGAGE Learning™

Australia • Brazil • Japan • Korea • Mexico • Singapore • Spain • United Kingdom • United States

Industrial Electrical Troubleshooting
Lynn Lundquist

Business Unit Director: Alar Elken
Executive Editor: Sandy Clark
Acquisitions Editor: Mark Huth
Editorial Assistant: Dawn Daugherty
Executive Marketing Manager: Maura Theriault
Channel Manager: Mona Caron
Marketing Coordinator: Kasey Young
Executive Production Manager: Mary Ellen Black
Project Editor: Barbara L. Diaz
Art Director: Rachel Baker

© 2000 Delmar, Cengage Learning

ALL RIGHTS RESERVED. No part of this work covered by the copyright herein may be reproduced, transmitted, stored, or used in any form or by any means graphic, electronic, or mechanical, including but not limited to photocopying, recording, scanning, digitizing, taping, Web distribution, information networks, or information storage and retrieval systems, except as permitted under Section 107 or 108 of the 1976 United States Copyright Act, without the prior written permission of the publisher.

For product information and technology assistance, contact us at
Cengage Learning Customer & Sales Support, 1-800-354-9706

For permission to use material from this text or product, submit all requests online at **www.cengage.com/permissions**.
Further permissions questions can be e-mailed to
permissionrequest.@cengage.com

Library of Congress Control Number: 99-049721

ISBN-13: 978-0-7668-0603-0

ISBN-10: 0-7668-0603-0

Delmar
Executive Woods
5 Maxwell Drive
Clifton Park, NY 12065
USA

Cengage Learning is a leading provider of customized learning solutions with office locations around the globe, including Singapore, the United Kingdom, Australia, Mexico, Brazil, and Japan. Locate your local office at **www.cengage.com/global**

Cengage Learning products are represented in Canada by Nelson ?Education, Ltd.

To learn more about Delmar, visit **www.cengage.com/delmar**

Purchase any of our products at your local bookstore or at our preferred online store **www.cengagebrain.com**

Notice to the Reader
Publisher does not warrant or guarantee any of the products described herein or perform any independent analysis in connection with any of the product information contained herein. Publisher does not assume, and expressly disclaims, any obligation to obtain and include information other than that provided to it by the manufacturer. The reader is expressly warned to consider and adopt all safety precautions that might be indicated by the activities described herein and to avoid all potential hazards. By following the instructions ?contained herein, the reader willingly assumes all risks in connection with such instructions. The publisher makes no representations or warranties of any kind, including but not limited to, the warranties of fitness for particular purpose or merchantability, nor are any such representations implied with respect to the material set forth herein, and the publisher takes no responsibility with respect to such material. The publisher shall not be liable for any special, consequential, or exemplary damages resulting, in whole or part, from the readers' use of, or reliance upon, this material.

Printed in the United States of America
5 6 7 8 9   25 24 23 22 21

Contents

Preface	ix
Brief Overview of the Chapters	xi
Acknowledgments	xiii
About the Author	xv

CHAPTER 1 Getting Started with Electrical Troubleshooting — 1
- Objectives — 1
- Introducing Troubleshooting — 1
- Looking Ahead — 4
- Troubleshooting Safety — 4
- Profitable Troubleshooting — 11
- Chapter Review — 12
- Thinking through the Text — 12

CHAPTER 2 Understanding Electrical Symbols — 13
- Objectives — 13
- Electrical Symbols — 13
- Diagram Numbering Systems — 25
- A Practical Application — 29
- Chapter Review — 30
- Thinking through the Text — 30

CHAPTER 3 Understanding Ladder Diagrams — 32
- Objectives — 32
- Introducing the Ladder Diagram — 32
- Reading the Ladder Diagram — 35
- Understanding the Ladder Diagram — 38
- Line Conductor Diagram — 39
- Wiring Diagrams — 41
- Chapter Review — 43
- Thinking through the Text — 43

CHAPTER 4 On-Line Troubleshooting Overview — 44
- Objectives — 44
- Example of On-Line Troubleshooting — 44
- Comparing Other Troubleshooting Methods — 50
- Chapter Review — 52
- Thinking through the Text — 52

CHAPTER 5 On-Line Troubleshooting Tools — 54
- Objectives — 54
- The Importance of Test Equipment — 54
- Contact and Non-Contact Measurements — 55
- Moving-Coil Meters — 57
- True RMS Meters — 66
- Choosing an On-Line Contact Meter — 67
- Non-Contact On-Line Troubleshooting — 70
- Troubleshooting with Specialized Continuity Testers — 73
- Testing Programmable Logic Controller Circuits — 74
- Chapter Review — 75
- Thinking through the Text — 75

CHAPTER 6 Collecting Information — 76
- Objectives — 76
- Stop, Look, and Listen — 76
- Troubleshooting Sequence — 77
- Preventive Information — 81
- Chapter Review — 82
- Thinking through the Text — 82

CHAPTER 7 Practical On-Line Troubleshooting — 84
Objectives — 84
Collecting Preliminary Information — 84
Preliminary On-Line Troubleshooting — 87
On-Line Troubleshooting — 90
Special Setup Troubleshooting — 93
Evaluating the Procedures — 96
Chapter Review — 97
Thinking through the Text — 98

CHAPTER 8 Troubleshooting with a Multimeter — 99
Objectives — 99
General Precautions — 99
The Basis of Multimeter Troubleshooting — 100
Specialized Multimeter Testing — 103
Evolution of the DMM — 106
Computer-Enhanced Multimeters — 108
Chapter Review — 112
Thinking through the Text — 113

CHAPTER 9 Specialized Tests and Equipment — 114
Objectives — 114
New Troubleshooting Possibilities — 114
Troubleshooting with Ammeters — 115
Troubleshooting Insulation Problems — 120
Troubleshooting Capacitor Problems — 128
Troubleshooting with Phase Meters — 130
Troubleshooting with Wire Identification Instruments — 132
Innovative Electrical Troubleshooting — 136
Troubleshooting with a Neon Lamp — 136
Chapter Review — 137
Thinking through the Text — 138

CHAPTER 10 Troubleshooting with Industrial Oscilloscopes — 139
Objectives — 139
Industrial-Rated Oscilloscopes — 139
Waveform Testing: General Diagnostics — 142
Trend Plot Testing: Evaluating Equipment over Time — 147
Chapter Review — 158
Thinking through the Text — 159

CHAPTER 11 Expanding On-Line Troubleshooting Applications — 160
Objectives — 160
Reading Complex Electrical Diagrams — 160
Reading Simple Wiring Diagrams — 161
Taking Your Troubleshooting Skill to the Plant — 165
Expanding Your Troubleshooting Capabilities — 171
Chapter Review — 171
Thinking through the Text — 172

CHAPTER 12 Broadening the Electrician's Horizons — 174
Objectives — 174
The Place for Broader Knowledge — 174
Choosing to Broaden your Knowledge — 176
Tools for Broadening your Knowledge — 179
A Concluding Thought — 181
Chapter Review — 181
Thinking through the Text — 182

CHAPTER 13 Troubleshooting Hydraulic and Pneumatic Systems — 183
Objectives — 183
Hydraulic Diagrams — 184
Hydraulic Symbols — 185
Troubleshooting with a Hydraulic Diagram — 188
Electrical Problems in the Hydraulic System — 190
Troubleshooting and Repairing Hydraulic Systems — 190
Pneumatic Symbols — 192
Chapter Review — 194
Thinking through the Text — 195

Appendix A — 197
Appendix B — 203
Key Terms — 207
Index — 217

DELMAR PUBLISHERS IS YOUR ELECTRICAL BOOK SOURCE!

In addition to the *Illustrated Guide to the National Electrical Code®*, Delmar Publishers offers a comprehensive selection of electrically oriented best-sellers and all new titles designed to bring you technically accurate, leading-edge information. *Electrical Wiring—Commercial*, and *Electrical Wiring—Industrial* may be of particular interest to you, as they complement the materials found in *Industrial Electrical Troubleshooting*.

NATIONAL ELECTRICAL CODE

National Electrical Code® 1999/NFPA
Revised every three years, the *National Electrical Code®* is the basis of all U.S. electrical codes.
Order # 0-8776-5432-8
Loose-leaf version in binder
Order # 0-8776-5433-6

National Electrical Code® Handbook 1999/NFPA
This essential resource pulls together all the extra facts, figures, and explanations you need to interpret the 1999 *NEC®*. It includes the entire text of the Code, plus expert commentary, real-world examples, diagrams, and illustrations that clarify requirements.
Order # 0-8776-5437-9

Illustrated Changes in the 1999 National Electrical Code®/O'Riley
This book provides an abundantly illustrated and easy-to-understand analysis of the changes made to the 1999 *NEC®*.
Order # 0-7668-0763-0

Understanding the National Electrical Code®, 3E/Holt
This book gives users at every level the ability to understand what the *NEC®* requires, and simplifies this sometimes intimidating and confusing code.
Order # 0-7668-0350-3

Illustrated Guide to the National Electrical Code®/Miller
Highly detailed illustrations offer insight into Code requirements, and are further enhanced through clearly written, concise blocks of text that can be read very quickly and understood with ease. Organized by classes of occupancy.
Order # 0-7668-0529-8

Interpreting the National Electrical Code®, 5E/Surbrook
This updated resource provides a process for understanding and applying the *National Electrical Code®* to electrical contracting, plan development, and review.
Order # 0-7668-0187-X

Electrical Grounding, 5E/O'Riley
Electrical Grounding is a highly illustrated, systematic approach for understanding grounding principles and their application to the 1999 *NEC®*.
Order # 0-7668-0486-0

ELECTRICAL WIRING

Electrical Raceways and Other Wiring Methods, 3E/Loyd
The most authoritative resource on metallic and nonmetallic raceways, provides users with a concise, easy-to-understand guide to the specific design criteria and wiring methods and materials required by the 1999 *NEC®*.
Order # 0-7668-0266-3

Electrical Wiring—Residential, 13E/Mullin
Now in full color! Users can learn all aspects of residential wiring and how to apply them to the wiring of a typical house from this, the most widely used residential wiring book in the country.
Softcover Order # 0-8273-8607-9
Hardcover Order # 0-8273-8610-9

House Wiring with the NEC®/Mullin
The focus of this new book is the applications of the *NEC®* to house wiring.
Order # 0-8273-8350-9

Electrical Wiring—Commercial, 10E/Mullin and Smith
Users can learn commercial wiring in accordance with the *NEC®* from this comprehensive guide to applying the newly revised 1999 *NEC®*.
Order # 0-7668-0179-9

Electrical Wiring—Industrial, 10E/Smith and Herman
This practical resource has users work their way through an entire industrial building—wiring the branch-circuits, feeders, service entrances, and many of the electrical appliances and sub-systems found in commercial buildings.
Order # 0-7668-0193-4

Cables and Wiring, 2E/AVO
This concise, easy-to-use book is your single-source guide to electrical cables—it's a "must-have" reference for journeyman electricians, contractors, inspectors, and designers.
Order # 0-7668-0270-1

ELECTRICAL MACHINES AND CONTROLS

Industrial Motor Control, 4E/Herman and Alerich
This newly revised and expanded book, now in full color, provides easy-to-follow instructions and essential information for controlling industrial motors. Also available are a new lab manual and an interactive CD-ROM.
Order # 0-8273-8640-0

Electric Motor Control, 6E/Alerich and Herman
Fully updated in this new sixth edition, this book has been a long-standing leader in the area of electric motor controls.
Order # 0-8273-8456-4

Introduction to Programmable Logic Controllers/Dunning
This book offers an introduction to Programmable Logic Controllers.
Order # 0-8273-7866-1

Technician's Guide to Programmable Controllers, 3E/Cox
Uses a plain, easy-to-understand approach and covers the basics of programmable controllers.
Order # 0-8273-6238-2

Programmable Controller Circuits/Bertrand
This book is a project manual designed to provide practical laboratory experience for one studying industrial controls.
Order # 0-8273-7066-0

Electronic Variable Speed Drives/Brumbach
Aimed squarely at maintenance and troubleshooting, *Electronic Variable Speed Drives* is the only book devoted exclusively to this topic.
Order # 0-8273-6937-9

Electrical Controls for Machines, 5E/Rexford
State-of-the-art process and machine control devices, circuits, and systems for all types of industries are explained in detail in this comprehensive resource.
Order # 0-8273-7644-8

Electrical Transformers and Rotating Machines/Herman
This new book is an excellent resource for electrical students and professionals in the electrical trade.
Order # 0-7668-0579-4

Delmar's Standard Guide to Transformers/Herman
Delmar's Standard Guide to Transformers was developed from the best-seller *Standard Textbook of Electricity* with expanded transformer coverage not found in any other book.
Order # 0-8273-7209-4

DATA AND VOICE COMMUNICATION CABLING AND FIBER OPTICS

Complete Guide to Fiber Optic Cable System Installation/Pearson
This book offers comprehensive, unbiased, state-of-the-art information and procedures for installing fiber optic cable systems.
Order # 0-8273-7318-X

Fiber Optics Technician's Manual/Hayes
Here's an indispensable tool for all technicians and electricians who need to learn about optimal fiber optic design and installation as well as the latest troubleshooting tips and techniques.
Order # 0-8273-7426-7

A Guide for Telecommunications Cable Splicing/Highhouse
A "how-to" guide for splicing all types of telecommunications cables.
Order # 0-8273-8066-6

Premises Cabling/Sterling
This reference is ideal for electricians, electrical contractors, and inspectors needing specific information on the principles of structured wiring systems.
Order # 0-8273-7244-2

ELECTRICAL THEORY

Delmar's Standard Textbook of Electricity, 2E/Herman
This exciting full-color book is the most comprehensive book on DC/AC circuits and machines for those learning the electrical trades.
Order # 0-8273-8550-1

Industrial Electricity, 6E/Nadon, Gelmine, and Brumbach
This revised, illustrated book offers broad coverage of the basics of electrical theory and industrial applications. It is perfect for those who wish to be industrial maintenance technicians.
Order # 0-7668-0101-2

EXAM PREPARATION

Electrician's Exam Preparation/Holt
This comprehensive exam prep guide includes all of the topics on both the master and journeyman electrician competency exams.
Order # 0-7668-0376-7

REFERENCE

ELECTRICAL REFERENCE SERIES

This series of technical reference books is written by experts and designed to provide the electrician, electrical contractor, industrial maintenance technician, and other electrical workers with a source of reference information about virtually all of the electrical topics that they encounter.

Electrician's Technical Reference—Motor Controls/Carpenter
Electrician's Technical Reference—Motor Controls is a source of comprehensive information on understanding the controls that start, stop, and regulate the speed of motors.
Order # 0-8273-8514-5

Electrician's Technical Reference—Motors/Carpenter
Electrician's Technical Reference—Motors builds an understanding of the operation, theory, and applications of motors.
Order # 0-8273-8513-7

Electrician's Technical Reference—Theory and Calculations/Herman
Electrician's Technical Reference—Theory and Calculations provides detailed examples of problem solving for different kinds of DC and AC circuits.
Order # 0-8273-7885-8

Electrician's Technical Reference—Transformers/Herman
Electrician's Technical Reference—Transformers focuses on the theoretical and practical aspects of single-phase and 3-phase transformers and transformer connections.
Order # 0-8273-8496-3

Electrician's Technical Reference—Hazardous Locations/Loyd
Electrician's Technical Reference—Hazardous Locations covers electrical wiring methods and basic electrical design considerations for hazardous locations.
Order # 0-8273-8380-0

Electrician's Technical Reference—Wiring Methods/Loyd
Electrician's Technical Reference—Wiring Methods covers electrical wiring methods and basic electrical design considerations for all locations, and shows how to provide efficient, safe, and economical applications of various types of available wiring methods.
Order # 0-8273-8379-7

Electrician's Technical Reference—Industrial Electronics/Herman
Electrician's Technical Reference—Industrial Electronics covers components most used in heavy industry, such as silicon control rectifiers, triacs, and more. It also includes examples of common rectifiers and phase shifting circuits.
Order # 0-7668-0347-3

RELATED TITLES

Common Sense Conduit Bending and Cable Tray Techniques/Simpson
Now geared especially for students, this manual remains the only complete treatment of the topic in the electrical field.
Order # 0-8273-7110-1

Practical Problems in Mathematics for Electricians, 5E/Herman
This book details the mathematics principles needed by electricians.
Order # 0-8273-6708-2

Electrical Estimating/Holt
This book provides a comprehensive look at how to estimate electrical wiring for residential and commercial buildings with extensive discussion of manual versus computer-assisted estimating.
Order # 0-8273-8100-X

Electrical Studies for Trades/Herman
Based on Delmar's *Standard Textbook of Electricity*, this new book provides non-electrical trades students with the basic information they need to understand electrical systems.
Order # 0-8273-7845-9

Preface

Industrial Electrical Troubleshooting will take the reader from an introduction to basic electrical diagram symbols to advanced machinery troubleshooting in an industrial plant. The book fully explains ladder logic electrical diagrams and their use in the troubleshooting process.

Throughout this book, the theme of reducing equipment downtime to minimize costly production loss is emphasized. This is primarily achieved by teaching the reader how to knowledgeably select and test in the most probable areas of circuit fault. The reader is then shown how to best use test instruments to quickly identify faulty electrical components.

Safety is stressed throughout the book. Both the hazard of electrical shock and the hazards inherent with moving machinery and stored energy are thoroughly discussed.

The effective plant electrician must be aware of much more than the electrical systems powering and controlling production machinery. Increasingly, he or she must understand the mechanical, hydraulic, and pneumatic components on the equipment being serviced. This book strongly emphasizes the need for knowledge and skill development in all aspects of industrial plant maintenance. The troubleshooting procedures demonstrated in this book are realistic and practical.

Industrial Electrical Troubleshooting will give the reader a broad exposure to the test instruments available today. Attention is directed toward effective and safe use of low-cost multimeters, proximity voltage testers, and clamp-on ammeters. In addition, a wide range of testing procedures are demonstrated using specialty instruments from megohmmeters to hand-held oscilloscopes to wire sorting instruments, and others. In each case, the testing procedures demonstrated are in the context of the plant electricians' everyday world.

Brief Overview of the Chapters

Chapter 1, *Getting Started with Electrical Troubleshooting,* defines electrical troubleshooting with an emphasis on reducing manufacturing plant downtime and lost production. Safety is stressed in this chapter.

Chapter 2, *Understanding Electrical Symbols,* explains the symbols and numbering systems used in present-day ladder diagrams.

Chapter 3, *Understanding Ladder Diagrams,* demonstrates the use and interpretation of symbols on the ladder diagram. Proper grounding and safety issues in control design are also discussed.

Chapter 4, *On-Line Troubleshooting Overview,* gives a step-by-step example with a simple electrical diagram to introduce basic troubleshooting. On-line troubleshooting is compared with other methods; the appropriate safety issues are discussed.

Chapter 5, *On-Line Troubleshooting Tools,* introduces basic troubleshooting tools. The distinction is made between contact (live voltage) and noncontact (proximity) testing. The basic moving-coil volt-ohmmeter, as well as digital multimeters, is explained. Suggestions are given for selecting test instruments for electrical troubleshooting.

Chapter 6, *Collecting Information,* encourages the electrician to question machine operators and perform necessary preliminary evaluations before attempting actual troubleshooting. A comprehensive troubleshooting sequence is suggested.

Chapter 7, *Practical On-Line Troubleshooting,* takes the electrician through a complete electrical troubleshooting problem. Complete electrical diagrams are used to illustrate each diagnostic step. The setup for two machine troubleshooting procedures are explained.

Chapter 8, *Troubleshooting with a Multimeter,* explains ladder diagram troubleshooting with a multimeter. Ample diagrams are used to show each test point on the circuit with an explanation of the expected meter settings and display values. The chapter concludes with an explanation of advanced DMMs and their use with personal computers.

Chapter 9, *Specialized Tests and Equipment,* demonstrates how equipment such as ammeters, megohmmeters, circuit tracers, and other specialized test equipment can be used in reducing troubleshooting time. However, the chapter ends by emphasizing the troubleshooter's skill rather than test instruments as the basis for effective troubleshooting.

Chapter 10, *Troubleshooting with Industrial Oscilloscopes,* introduces the reader to test equipment and procedures that are increasingly becoming everyday experiences for the industrial electrician. Interesting examples of the oscilloscope's use in an industrial plant demonstrate its usefulness as a troubleshooting tool.

Chapter 11, *Expanding On-Line Troubleshooting Applications,* builds on the foundation of basic symbols and diagrams to demonstrate the ease with which complex ladder diagrams can be used. Pictorial wiring diagrams are also explained. The chapter ends with a discussion of fuses and their role in protecting equipment.

Chapter 12, *Broadening Your Horizons,* recognizes that good troubleshooting comes from a broad familiarity with technical and practical information. The electrician who aspires to become a good troubleshooter is encouraged to pursue continued professional growth.

Chapter 13, *Troubleshooting Hydraulic and Pneumatic Systems,* encourages the electrician to become knowledgeable about the hydraulic and pneumatic systems being maintained. A basic understanding of these systems' operation will aid in effective troubleshooting, thus reducing production downtime.

Appendix A is a complete electrical diagram for an injection molding machine. Many of the circuit diagrams throughout the book come from this schematic.

Appendix B is the complete hydraulic diagram for this same injection molding machine. This diagram is used for the hydraulic system examples in the last chapter.

Acknowledgments

This book could not exist without the many manufacturers who have developed and produced test instruments for the electrician's use in the field. The electrician is seldom aware of his or her dependence on the research, development, manufacture, and marketing of both the tools (including test instruments) and materials used in the electrical trade. We would not have our jobs if the manufacturing segment of the electrical industry could not make a profit. For this reason, I have identified good test equipment by manufacturer's name throughout the book. Regrettably, there are many good test instruments on the market that I have not been able to test or name.

Marketing and technical sales representatives of various companies have been my primary contacts. To these companies and individuals I owe my thanks for their help in making this book possible: A. W. Sperry Instruments, Inc., Kevin Basso; AVO International (Biddle), Jeffrey Jowett and Peg Houck; Cutler-Hammer, Paul Handle and Debbie Kasprzak; Greenlee Textron, Sandra Turner; IAEI, Phil Cox and Kathryn Ingley; Kawaguchi, Inc., Patrick Miura; Lincoln Electric Company, Dick Sabo; Littelfuse, Inc., Dan Stanek; Master Publishing, Inc., Peter Trotter; Miller Welding Equipment Company, Mike Pankratz; Panduit Corp, David Carson; Siemens Electromechanical Components Inc., Karl Grubb and Cathy Mead; Square D Company, Cynthia Corbett and Lisa Ratchford; Tektronix, Inc., K. L. Yeung, Marsha Bush, and Beth Daniels; Tif Instruments, Inc., Adam Seymour; Universal Enterprises, Inc. (UEI), Lisa Walker and Bill Raymond; and Vickers, Inc., Frank Garner.

As you read *Industrial Electrical Troubleshooting,* you will see FLUKE instruments heavily represented. There is a reason. No company worked as hard as FLUKE Corporation at making their equipment and their technical personnel available to me. I am immensely impressed as both a writer and an electrician with the quality of their product and their commitment to service. Thank you to the few whom I can name and the many I cannot: Debby Coyne, Julie Kuntz, David Pereles, and Edward Shen.

Thanks is also due to my wife, Gail, for her patience and help in the writing project.

The author and Delmar would like to extend thanks to the members of the review panel who provided detailed reviews of the manuscript. The contributions and suggestions of the following individuals are greatly appreciated.

Stephen Roggy
Greenville Technical College
Greenville, SC

Larry Pogoler
Los Angeles Trade-Technical College
Los Angeles, CA

Thomas Pickren
Albany Technical Institute
Albany, GA

Kevin Weigman
Northeast Wisconsin Technical College
Green Bay, WI

About the Author

Lynn Lundquist is a veteran troubleshooter, licensed plant supervisor, and general journeyman electrician. He was the general maintenance manager in plastic injection molding and die casting shops for eight years. He has worked in research and development and supervised the electrical department for a pipe manufacturing facility. In addition, he is an experienced machinist, certified welder, and metal fabricator. He has designed and built automated equipment and holds twenty-three issued U.S. patents. In addition, he has two patents pending in the field of electrical test instruments. Mr. Lundquist has both a B.A. and a Master's degree in education and has published five books.

CHAPTER 1

Getting Started with Electrical Troubleshooting

OBJECTIVES

After completing this chapter, you should be able to:
- Describe the steps to effective industrial plant troubleshooting.
- Understand the value of effective troubleshooting to both yourself as an industrial plant electrician and to your employer's financial profitability.
- Define the term *on-line electrical troubleshooting*.
- Understand the current safety nomenclature used to classify electrical test instruments.
- Identify and use the safety practices that are mandatory for long-term effective troubleshooting.

INTRODUCING TROUBLESHOOTING

Were you able to significantly reduce the cost of lost production during your plant's last electrical emergency? If your answer is "yes," the chances of better pay and job advancement are in your favor.

No skill will better establish your value as a good maintenance electrician than your speed in getting equipment back into production after an equipment failure. Your value is even greater if you are able to complete your troubleshooting diagnosis while preventing the shutdown altogether. Effective troubleshooting skills should mean greater profitability for your employer and greater job security and pay for you.

In your last electrical emergency it may have taken you more time to diagnose the cause of the shutdown than it did to repair it. That is often the case. Because locating electrical problems is a major part of machine **downtime**, this book will emphasize on-line troubleshooting techniques that will speed the process of locating a **fault**. *The objective of this book is to help you reduce the time it takes you to find an electrical problem in the equipment you are testing.*

Defining Electrical Troubleshooting

After an electrical problem has caused the complete failure of major equipment, identifying components that need replacement is often simple. However, waiting until there is extensive damage so that an entire system can be replaced is an inefficient and expensive way to attempt troubleshooting. It also

2 CHAPTER 1 Getting Started with Electrical Troubleshooting

inevitably results in considerable lost production time. In this text, we want to both define and practice a troubleshooting technique that is efficient and keeps production equipment running. Consequently, troubleshooting is defined as *locating a single component failure* before extensive damage is done.

The idea of a *single* component failure is the *first* component that fails or is operating abnormally. When allowed to continue operating at less than optimum performance, this *first* weak component will place the rest of the system in jeopardy. There may also be a cause that is more fundamental than an electrical or mechanical component failure. There may be a root cause that extends beyond an actual component. A shop rag or carelessly dropped candy wrapper that has been drawn into an air cooling passage may be the root cause of equipment failure. All of these possibilities are included in our definition, which requires *locating a single component failure before extensive damage is done*.

The electrician in the next department does not understand this definition of troubleshooting. Three days ago—after airing out a compressor room—it was obvious that the motor was "smoked." "That," our electrician friend maintains, "is the simple way to troubleshoot—just look for the burned-out stuff!" Of course, in addition to the motor, other expensive equipment also had to be replaced: the magnetic **contactor**, the pressure switch, a timer, and other miscellaneous components. However, the electrician did not mention in his requisition for new equipment that for two days before the motor finally quit, the **motor overloads** had to be continually reset and a sticking head **pressure-relief valve** as shown in Figure 1–1 had been tampered with to keep the compressor running.

Effective troubleshooting will find the first electrical or mechanical malfunction of any component part on which the normal operation of that electrical system is dependent. By doing that successfully, downtime is reduced, equipment damage is limited, total repair costs are minimized, and you, the electrician, look good!

A brief discussion of the above definition is worthwhile. Good troubleshooting usually follows a logical progression.

1. *Good troubleshooting will find the first electrical or mechanical malfunction of any component*

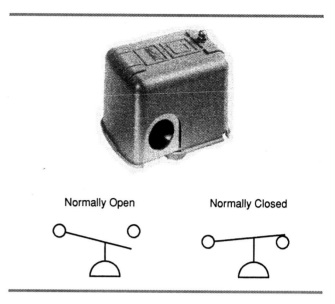

Figure 1-1 An example of an air compressor pressure switch having both electrical contents and a compressor head pressure-relief valve. The symbols of normally open and normally closed pressure switch contacts are shown. Courtesy of Square D Company.

part. This should be obvious, but in practice it is often violated. Troubleshooting should start at the first indication of an electrical or mechanical malfunction. It should then begin at the level of finding the first component part that is not operating properly. Too often, erratic running is allowed to continue until major electrical damage is done. An example was given of the electrician who finally replaced an entire motor and starting system. If, on the other hand, the first electrical or mechanical malfunction of a component part had been isolated, nothing more than the faulty head pressure-relief valve would have been replaced. Not only would the electrician have saved a great deal of company money if the troubleshooting had been done early, but if our simple on-line troubleshooting procedure had been used, the total troubleshooting time could have been reduced to 15 or 20 minutes.

We have not limited our definition of troubleshooting to the first *electrical* malfunction. It may be a mechanical malfunction that is causing electrical equipment to shut down or operate improperly. A plant maintenance electrician is often the first person called when equipment will not operate. Good skills in diagnosing mechanical problems are just as important to an effective troubleshooting electrician as electrical skills. Worn pillow block bearings on a belt-driven shaft can trip motor overloads. A damaged bracket on an access door may prevent a **limit switch** from functioning. A low oil level or high crankcase temperature may prevent an air compressor from starting. In many industrial plants, the maintenance electrician may actually be involved in mechanical repairs. On the other hand, if these repairs will be done by millwrights, the total repair time will still be reduced if the electrician does not waste time looking for nonexistent electrical problems. It is embarrassing to be in the middle of troubleshooting a **direct current** drive system when a millwright asks, "Did you notice that the auger bearing has failed again?" Throughout this book, we will repeatedly return to the electrician's involvement with the mechanical side of industrial plant maintenance.

2. *Good troubleshooting will find the first electrical or mechanical malfunction on which the normal operation of that system is dependent.* Good troubleshooting views a modern electrical installation as a system. Proper maintenance and good troubleshooting must recognize that even the most complex system is dependent on a series of discrete electrical and mechanical components. A single discrete electrical device such as a compressor head pressure-relief valve is as crucial to the overall operation of the compressor system as the 15–horsepower compressor motor. Too often, however, the small components are overlooked until the entire system is in jeopardy. On-line troubleshooting, by its very nature, forces you to think about every part of an electrical system. As you maintain equipment, you should not become too busy to notice the small and inexpensive electrical and mechanical items on which the entire system is dependent.

Defining On-line Troubleshooting

Before going further, we need to define *on-line troubleshooting*. On-line troubleshooting incorporates the two points of good troubleshooting previously mentioned. If you look closely at the definition, however, you will realize that nothing has been said about the troubleshooting procedure.

Chapter 4 gives a complete introduction to the on-line troubleshooting technique. For now, you need a very brief explanation to help you understand the concept as you read the next few chapters.

There are many different testing procedures that can be used in electrical troubleshooting. Some are done with the electrical power off, and some with the power on. In this book, we emphasize a number of procedures where the testing is done on a live **control circuit**—that is the meaning of "on-line." If one takes the necessary safety precautions, there are some advantages to on-line electrical testing. The primary advantage is in testing a circuit as it normally operates. Many times when electricians de-energize a faulty circuit, the problem disappears. They may then spend a great deal of time testing for a problem that no longer exists.

In on-line troubleshooting you will learn to do a series of tests on a circuit that is energized but is not operating normally. You will learn procedures for isolating the point in that circuit at which some electrical component has failed. You will also be introduced to a number of test instruments that will make your work possible. If you are familiar with electrical testing equipment, you know that standard **volt-ohmmeters (VOMs)** are not used for resistance measurements while the circuit is energized. There are meters presently available that can be used for these kinds of "live" measurements. It is because of these new meters that on-line techniques are possible.

In on-line troubleshooting, you will be testing for **continuity** on a live circuit as illustrated in Figure 1–2. Continuity means that the circuit is complete. You will trace the live circuit to the break or open condition that is causing the circuit to malfunction. At times, you will be doing your testing on both energized and operating circuits.

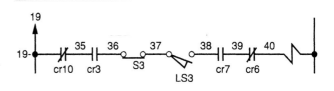

Figure 1-2 The solenoid at the right of the circuit will not operate because LS3 is open. Placing meter leads on wires 37 and 38 will show that the circuit is open. This test can be done on the continuity setting of a meter capable of reading both continuity and voltage on an energized control circuit.

Building a Troubleshooting Foundation

Electrical troubleshooting covers a broad spectrum. Initially, we encounter electrical circuits consisting of **electromechanical** equipment that opens and closes **electrical contacts**. In this book, we are emphasizing circuits that control individual **relays**. The individual relays in a **control panel** become an integrated **relay logic** system that performs decision-making functions with mechanical relays. The decision is either "yes" (because the relay is conductive) or "no" (because the relay is open). This relay logic regulates the **solenoids**, motor **starters**, and any other equipment at the end of each control circuit path. In turn, these solenoids and motor starters start and stop the motors and other power equipment, which is the ultimate object of industrial equipment control. Though all of this equipment has been greatly improved over the years, its basic function remains essentially unchanged from that of its predecessors of 30 or more years ago.

With the advent of **solid-state** equipment, followed by the development of integrated circuits, a new generation of control and output equipment began to emerge during the 1960s. In the initial stages, this consisted of replacing kind-for-kind. A mechanical clock motor timer was replaced with a solid-state timer, or a direct current motor-generator set was replaced with a solid-state direct current drive. In most cases, however, the links between the solid-state equipment continued to be mechanical switching and relays.

Today, much of the control of industrial process has been taken over by electronic equipment. **Programmable logic controllers** (PLCs) now replace many of the earlier relay logic panels. So-called **smart meters** process production information and are capable of sophisticated switching functions. A new generation of **serial addressable output** devices are now on the market that communicate among themselves and a central computer processor in what is known as *DeviceNet*. A well-qualified industrial electrician must become familiar with this new equipment.

In this book, however, we will focus much of our attention on troubleshooting **hard-wired** circuits. An understanding of these physical circuits is a prerequisite to successful work with sophisticated controls of any kind. Furthermore, the **termination point** for the most advanced controls is the very equipment we are learning to troubleshoot in this text. Effective troubleshooting, as described in this book, will never be obsolete. It is the foundation on which an industrial electrician's work is built.

LOOKING AHEAD

As you read this book, your goal should be to become a better troubleshooter. On-line troubleshooting techniques will make you faster in locating electrical problems. In some cases, this technique will virtually eliminate machine downtime during the troubleshooting process. It is also important for you to understand related nonelectrical systems that have a bearing on your effectiveness in solving electrical problems. This will include a basic understanding of **hydraulic** and **pneumatic systems** and the ability to read the appropriate prints. There is great value in a maintenance electrician who has the ability to troubleshoot an electrical system on hydraulic or pneumatic equipment. To successfully do that, the electrician will need to understand the basic functions of the system itself. The electrician may frequently design (and even build) pneumatic systems.

Throughout this book you will work with typical electrical malfunctions. You will troubleshoot a plastic injection molding machine using the actual **diagrams** and realistic electrical problems. If you understand injection molding equipment, it will add interest to the examples. However, no prior knowledge is necessary, since the diagrams you will be using are standardized for all industrial equipment. You will be using a basic **ladder diagram** that is similar for any relay-controlled machine.

A complete ladder diagram is given in Appendix A. All of the electrical circuit examples will come from this one diagram. To visualize the relationships among the machine's electrical circuits, it will be helpful for you to spend some time looking at the entire diagram.

TROUBLESHOOTING SAFETY

The importance of safe troubleshooting procedures cannot be stressed enough. Safety for both personnel and equipment must be the first priority in any troubleshooting job. The need for safety does not eliminate well-planned shortcuts or time-saving procedures, but speed cannot become such a dominant goal that prudence is set aside.

The logical place to start a discussion of safety is with work on live circuits. This is particularly true since this book advocates a test procedure that uses energized equipment. The admonition to shut everything down before testing has long been stated as an ideal. However, for most testing, some live-circuit tests must be done. The most frequently used test meter in the electrician's toolbox is a voltmeter. This obviously is not a piece of test equipment intended for disconnected circuits. Live-circuit testing is very much a part of the electrician's world.

The question, then, is not simply whether the circuit is energized. The greatest concern is the care taken in using whatever procedure is required. Any troubleshooting procedure is dangerous when safety precautions are ignored. On the other hand, on-line troubleshooting with a live circuit can be done safely. The real issue is the attention given to safety by the electrician doing the work.

In the past, initial troubleshooting was done with electrical circuits deenergized. However, when potential sources of circuit failure were located, the circuit was often reenergized with **jumper** wiring used to replace presumably faulty components. Needless to say, the potential of hazard is significantly greater using jumpers to start equipment than is encountered when merely reading a meter for either a voltage or continuity value. When properly used, on-line troubleshooting is considerably safer than a procedure that requires temporary wiring to be used to bypass electrical contacts on an operating circuit.

Additionally, you can actually introduce risk when working on presumably deenergized circuits because of multiple circuits in a work area. The control circuit you are working on (usually 120 **volts** or less) may well be dead, yet there may be live 240– or 480–volt circuits running through relays in the same panel that have not been deenergized. Those isolated, energized circuits have an uncanny way of finding your screwdrivers and pliers—if not you— because you were working as though the entire panel were dead. The *National Electrical Code® (NEC®)* requires that a single **disconnect** be used for both the motor and the reduced voltage control. Generally, this means that the disconnect will deenergize the entire panel. However, allowable Code exceptions and old or improperly wired equipment can introduce live equipment hazards.

Even though you are working on a live circuit, safety can be enhanced by either the way a meter is used or by the meter function chosen. The motor disconnect shown in Figure 1–3 controls a water pump. In this particular installation, a fusible disconnect, the motor contactor, and the pressure and tank air-cushioning controls are inside a plant building. The well and pump motor are located in a separate shed approximately 100 feet away. The pump motor has an adjacent **fused disconnect** that can be used as a **lockout**. Though two lockouts are desirable, a single lockout would satisfy *NEC®* requirements. The two sets of **fuses**, however, are redundant.

When testing the pump motor circuit from the motor contactor, it is obvious that the mere presence of a voltage on the load side of the contactor does not indicate that the motor is operating normally. If two fuses in the remote disconnect were blown, the contactor would show a full 480 volts at the **load terminals**, but the motor would not be running. We can thus use any one of three testing procedures. Each will have differing degrees of potential hazard as well as give different information regarding the pump motor's performance.

Figure 1–3A shows a voltage test. With this test you can confirm that the primary fuses are conductive and that the contactor is properly closed. The meter is being used properly, and there is minimum hazard potential. Because the meter is in electrical contact with the circuit, this is a **contact test**. However, this test cannot verify that the motor is running.

Figure 1–3C shows a **proximity tester** being used on the load side of the contactor. This test has the advantage of further reducing the hazard potential because there is no contact with live **conductors**. It is a **non-contact test**. This is an excellent test procedure before performing work on the circuit after locking out the disconnect. On the other hand, a proximity test is not reliable as a test of the **line voltage** to the motor. You could actually have two blown fuses in the main disconnect and still have a "hot" reading with a proximity meter because the leads are **common** in the motor. *Induction* from the parallel

6 CHAPTER 1 Getting Started with Electrical Troubleshooting

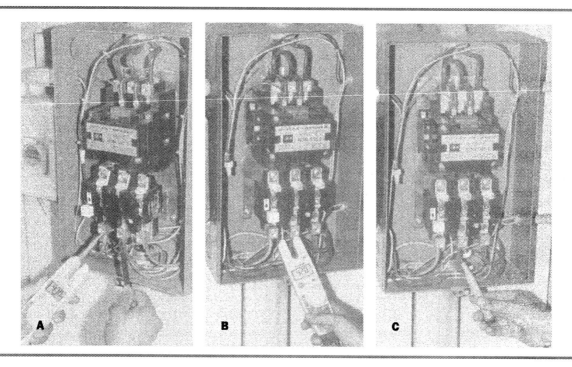

Figure 1-3 Safety may be enhanced by proper choice or use of the meter. (A) A properly conducted voltage test is considered safe, but the meter is in contact with the line voltage. (B) A current test gives an indication of motor operation with the added benefit that it is not in contact with line voltage. (C) A proximity tester indicates the presence of an energized conductor, yet it is not in contact with the line voltage. *The T5-1000 courtesy of FLUKE Corporation.*

lines would also cause the proximity meter to indicate an energized line.

There is a third option shown in Figure 1–3B that has the dual advantage of the reduced hazard of a non-contact test and the best indication of motor performance. A FLUKE T5–1000 is being used to read **current**. If all three legs are drawing a balanced current load, the contactor, all fuses, and the motor itself have been verified as performing satisfactorily. At the same time, you have reduced the hazard potential by using a test procedure that is not in contact with live voltage. In many instances, properly choosing the test procedure (and/or the best suited test meter) can enhance the safety of electrical testing.

Some testing on live circuits will always be necessary. Choosing testing procedures and testing equipment carefully can help reduce the risk of an accident. Nonetheless, when you have a choice, work on a deenergized circuit. Unnecessarily sticking a screwdriver into an energized panel circuit to fiddle with **device terminal** screws or push wires around is asking for trouble. Make it a practice to take the extra time necessary to turn off the power when it is possible to do so. Always follow lockout/**tagout** procedures when they apply, using appropriate methods as illustrated in Figure 1–4. In addition, be aware of both your state and your individual employer's lockout/tagout requirements. In some cases they may have more stringent standards than the **Occupational Safety and Health Administration (OSHA)**. Under no circumstances should work be done on power leads to motors, **transformers**, or similar equipment without complete lockout procedures in place. Equally, any maintenance work on a control circuit that involves disconnecting wires should be done on **secured**, deenergized circuits as required by the OSHA lockout/tagout rules given in Figure 1–5.

On-line troubleshooting does not introduce any new hazards to electricians. For years we have all used troubleshooting procedures on live circuits, though we have been measuring different values than we will be with this method. The safety requirements, however, *have not substantially changed*.

Figure 1-4 Lockout/tagout procedures must be followed. (A) Multiple disconnects may be locked out with a Multiple Lockout Device. (B) Lockout/tagout compliance may involve a single circuit. (C) The electrician must secure all energy sources if the electrical work creates nonelectrical hazards. *Reprinted with permission of PANDUIT Corp.*

Test Meter Ratings

As distribution systems and loads become more complex, the possibilities of **transient** overvoltages increase. Motors, **capacitors** and power conversion equipment such as **variable speed drives** can generate voltage transients. Lightning strikes on outdoor transmission lines can also cause extremely hazardous high-energy transients. If you are taking measurements

---

h3/1910.333—Selection and use of work practices
(a)(1)
"Deenergized parts." Live parts to which an employee may be exposed shall be deenergized before the employee works on or near them, unless the employer can demonstrate that deenergizing introduces additional or increased hazards or is infeasible due to equipment design or operational limitations...
  Note 2: Examples of work that may be performed on or near energized circuit parts because of infeasibility due to equipment design or operational limitations include testing of electric circuits that can only be performed with the circuit energized...
(a)(2)
"Energized parts." If the exposed live parts are not deenergized (i.e., for reasons of increased or additional hazards or infeasibility), other safety-related work practices shall be used to protect employees who may be exposed to the electrical hazards involved... The work practices that are used shall be suitable for the conditions under which the work is to be performed and for the voltage level of the exposed electric conductors or circuit parts.
(b)(2)
"Lockout and Tagging." While any employee is exposed to contact with parts of fixed electric equipment or circuits which have been deenergized, the circuits energizing the parts shall be locked out or tagged or both...
(b)(2)(ii)(B)
The circuits and equipment to be worked on shall be disconnected from all electric energy sources. Control circuit devices, such as push buttons, selector switches, and interlocks, may not be used as the sole means for deenergizing circuits or equipment. Interlocks for electric equipment may not be used as a substitute for lockout and tagging procedures.
(b)(2)(ii)(D)
Stored non-electrical energy in devices that could reenergize electric circuits parts shall be blocked or relieved to the extent that the circuit parts could not be accidentally energized by the device.
(b)(2)(iii)(A)
A lock and tag shall be placed on each disconnecting means used to deenergize circuits and equipment on which work is be performed... The lock shall be attached so as to prevent persons from operating the disconnecting means unless they resort to undue force or the use of tools.
(b)(2)(iii)(C)
If a lock cannot be applied, or if the employer can demonstrate that tagging procedures will provide a level of safety equivalent to that obtained by the use of a lock, a tag may be used without a lock.

Figure 1-5 Occupational Safety and Health Administration (OSHA) rules for logout/tagout procedures from Standards Number 1910.333. *Code of Federal Regulations.*

on electrical systems, these transients are "invisible" and largely unavoidable hazards. They occur regularly on low-voltage power circuits, and can reach peak values in the many thousands of volts. In these cases, you are dependent for protection on the safety margin already built into your meter. *The* **voltage rating** *alone will not tell you how well a meter was designed to survive high transient impulses.*

We need to briefly look at a transient. A transient is a rapidly changing voltage value imposed on the supply line, generally from an outside source. Typically, transients are of a very short time duration, but reach exceedingly high values. Figure 1–6 shows the transient produced by switching a relay coil. As the **electromagnetic field** surrounding the coil suddenly changes, it induces a high voltage value. In the case of this test, the voltage on a 120–volt relay coil produced multiple voltage spikes of approximately 850 volts. The short duration of this transient would have no measurable effect on something like a motor winding. On the other hand, a test meter or equipment with solid-state circuitry is another matter. Solid-state components are built with an upper limit of allowable insulation value. Beyond that limit, the insulation can be punctured and the circuit will be ruined. For the same reason, transients can pose extreme hazards to personnel. Even though the voltage spikes are generally of a very short duration, they are, nonetheless, critically high.

Now, imagine that you are using a meter designed for a maximum of 1,000 volts. While you are taking a measurement on a 480–volt circuit, say a transient of 2,500 volts comes through the conductors you are testing. That 2,500 volts may instantly break down the insulation in the solid-state circuit. Say for instance that the meter's insulation has now been damaged enough by this single voltage spike so that it can now tolerate only 80 volts as the 60 **hertz (Hz) cycle** drops to zero. But you, of course, are holding the test leads on a 480–volt motor contactor. In 1/120 of a second the sine wave again goes from zero to peak voltage. You are now holding a meter across 480 volts that develops a direct short. In another 1/120 of a second the meter will begin to draw the full current of the directly shorted circuit through the leads you are holding in your hands. What will happen to the meter? What could happen to you if

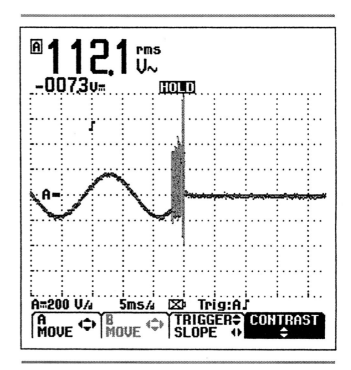

Figure 1-6 An example of transient voltage produced by a small 120-volt relay coil when switched. The maximum voltage exceeded 800 volts. The transient in this figure was obtained on a stable 120-volt power source applying a properly functioning coil. The transient was a normal random occurrence each time the coil was switched.

you are holding the meter in your bare hands and are not wearing eye protection?

The real issue for **multimeter** circuit protection is not just the maximum **steady-state voltage** range but a combination of both steady-state and transient overvoltage **withstand capability**. Transient protection is vital. When transients ride on high-energy circuits, they tend to be more dangerous because these circuits can deliver large currents. If a transient causes an arc-over, the high current can sustain the arc, producing a **plasma breakdown** or explosion, which occurs when the surrounding air becomes **ionized** and conductive. The result is an arc blast, a disastrous event that causes more electrical injuries every year than the better-known hazard of electric shock.

In order to more adequately protect users of electrical test equipment, the **International Electrotechnical Commission (IEC)** has issued a new IEC 1010 standard that offers a significantly higher level of safety. This standard recognizes four Overvoltage Installation Categories, *shown in Figure 1–7*. These

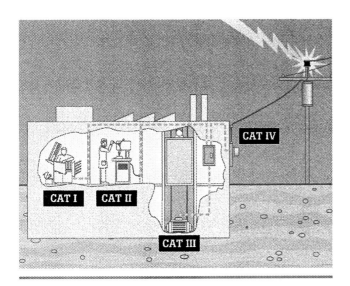

Figure 1-7 The International Electrotechnical Commission (IEC) has established a standard indicating the transient overvoltage withstand rating of a meter. Higher CAT numbers (I, II, III, and IV) identify an environment with higher power availability and energy transients. *Country of FLUKE Corporation.*

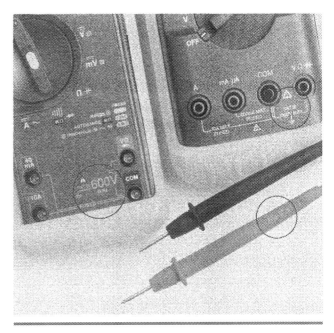

Figure 1-8 Better electrical meters now indicate both the steady-state and the CAT withstand voltage levels. Meter leads must be similarly protected. *Courtesy of FLUKE Corporation.*

are identified as Categories I through IV (CAT I, CAT II, CAT III, and CAT IV), in descending order from the highest transient potential:

- CAT IV identifies the **three-phase** utility connection and any conductors located outdoors.
- CAT III identifies all three-phase distribution and **single-phase** commercial lighting.
- CAT II identifies all single-phase receptacle-connected loads.
- CAT I identifies electronic equipment that includes high-voltage transformers within the equipment itself. The primary hazard in CAT I and CAT II is electric shock rather than transients and arc blast.

Reducing hazards associated with electrical testing involves more than merely using a meter properly. Manufacturers, testing laboratories, and standards committees are all working toward protecting the end user from hazards that extend beyond the test site itself. The very real danger of transients and associated dangers is now a part of better meter design. The industrial electrician who is working safely will also be concerned with the combined withstand rat-

ings of the meter. Typical meter identification is shown in Figure 1–8.*

General Electrical Safety Precautions

To maintain a high level of electrical safety while working, there are a number of precautions you should follow. This is, of course, only a start. Your individual work area will have its own unique safety requirements. The voltage levels and the types of equipment you are working on in your own plant will introduce conditions that will make demands beyond those given here. We are assuming that voltages will not exceed what the *National Electrical Code*® classifies as "600 volts or less." However, the following are given as essential basics:

1. *On-line troubleshooting is recommended only for control voltages of 120 volts or less.* This is never a technique for higher-voltage circuits, nor is it to be used on motors or any other high-

---

* The information in this section is taken from *ABCs of Multimeter Safety*, published by FLUKE Corporation.

current applications. In later chapters in the book, line voltage tests will be described. Additional safety precautions must always be enforced when working on line voltages.

2. *Always remember that even with a 120–volt control system, there may be higher voltages in the panel.* Always work so that you are clear of any of the higher voltages.

3. *Avoid unnecessary exposure to potential shock.* Use your meter carefully when touching **exposed current-carrying** terminals. *Do not move wires to gain better access when the power is on.* Above all, *shut the power off before starting any required repair work.*

4. *Wear proper protective gear.* Work should never be done around electrical equipment without proper eye protection. Arcing from shorted lines or outside transients exceeds the fastest human reflex time. Directly shorted conductors can throw molten metal into an electrician's face. Other protective gear, including gloves, insulating floor mats, **grounding jumpers,** and burn-resistant clothing, should be used where appropriate as illustrated in Figure 1–9. When working above certain heights, both OSHA and good judgment require safety harnesses. However, because of the involuntary reflex reaction to even low-voltage shock, the careful electrician may elect to use a safety harness on lower heights or around moving equipment.

5. *When working in a live panel, make it a practice to avoid leaning on or touching* **grounded** *areas.* In other words, do not lean against the panel or hold on to the door while you are checking circuits. Tie panel doors open when you are working inside the panel. A door swinging shut (or getting bumped) can bring the back of energized controls in contact with a working electrician. The reflex could result in even more serious electrical shock inside the panel.

6. *Use all test equipment safely and according to the manufacturer's recommendations.* Do not use it for higher voltages or in testing conditions for which it was not designed. If your meter is not **autoranging**, always start on the highest setting to get a low scale reading on the initial test; this will protect you and the meter in case the

Figure 1-9 Use proper personal safety equipment commensurate with the potential hazard. Always wear eye protection when working close to live circuits. *Courtesy of* Fluke *Corporation.*

voltage (or amperage) is higher than anticipated. When working on circuits subject to transient overvoltages, hang the meter on an appropriate support rather than holding it in your hand.

7. *Use good housekeeping procedures inside a panel.* Keep tools, wire spools, and all other odds and ends outside the panel when they are not in use. The inside of the panel is not a workbench. Keep it clear!

8. *Use good electrical practices even in temporary wiring for testing.* At times you may need to make alternate connections, but make them secure enough so that they are not in themselves an electrical hazard.

9. *If planning to work on a circuit as though it were deenergized, make certain that it is truly locked out.* Presumption at this point can kill you. It is a good practice to take a meter reading before starting work on a deenergized circuit. It is also a good practice to confirm the reading before relying on it. Proximity testers can easily be verified

by turning them on and tapping the probe vigorously against the hand. The capacitance in your hand is generally sufficient to create a signal. *Without releasing the "on" switch or clip,* place the probe next to the wires of the circuit to be contacted. Be aware that these probes may sometimes falsely indicate energized conductors because of induced fields from parallel wiring. When this happens, final verification must be made with a standard volt meter.

10. *Verify the state of all mechanical equipment.* Do not overlook mechanical equipment that will become involved during testing. This may require disconnecting linkages, clearing an area for movement, or simply verifying that equipment and personnel are out of harm's way.

11. *Pay attention to any condition that is abnormal or adds additional risk potential.* Industrial electricians will often encounter hazardous conditions that are transitory. That is, they are not normally a part of the maintenance setting for a particular piece of machinery. There are many examples, such as water where it is not normally encountered, broken support structures on a machine in the area of the electrical fault, spilled hydraulic fluid that makes the work area more dangerous, a heavy concentration of fumes because ventilation equipment failed, and the like. These hazards must be evaluated—and, in some cases, removed—before work is started.

Over and above all of the lists that could be made for safety practices, however, there is a single admonition that will produce the safest working conditions. *Use good judgment.* The electrician should have the best understanding of what a particular job entails. With that as background, and with a proper understanding of the inherent hazards in electrical work, you should then be able to decide the limits of acceptable safety practice as you proceed.

PROFITABLE TROUBLESHOOTING

If you master the simple technique of on-line troubleshooting as developed in this book, you will seldom spend more than 20 or 30 minutes isolating an electrical malfunction in relay or mechanical contact-actuated equipment. What is more, you may be able to do the entire job up to the final confirmation of the faulty component from the electrical panel. You will not be testing individual components and switches at their location. You will learn how to take almost all the readings from inside the panel, allowing you to isolate faulty switches or electrical equipment (solenoids, etc.) from the **terminal block** of the panel.

Of course, you should have a basic understanding of the equipment you are troubleshooting. In Chapter 6 you will be told how to work through the problem if you do *not* know the equipment. You should also have a complete ladder diagram of the equipment. In Chapter 11, you will learn how to work without a ladder diagram if it is not available. It should be obvious that the machine should be represented by an up-to-date ladder diagram. Nonetheless, be prepared. You will open the panel some day and find the rewiring work of an "old-timer" who left jumper wires throughout the panel without making a single notation of any of the changes on the electrical diagram! All of the wires should be properly marked with **wire numbers** in accordance with the ladder diagram notations. In Chapter 11, you will also learn how to identify a wire which has lost its number.

A DAY AT THE PLANT

*Though it is embarrassing, a personal mistake serves as a good reminder of potential hazard when mechanical functions are overlooked. I was test running a 5–horsepower, 3,600–rpm vertical shaft motor used for a direct-coupled wire cup brush. I removed the brush holder from the shaft and cleared the area before starting the motor. As the motor was coasting to a stop after the first **jog**, an observant senior maintenance supervisor pointed out that I had forgotten to remove the key from the shaft! A key thrown from the motor shaft could easily take out an eye.*

Finally, you should have the appropriate test equipment to do the job. There will be a complete description of the equipment needed for on-line troubleshooting in Chapter 5. Later chapters will introduce additional equipment and testing procedures that will help you do a better and faster job. What you need will be neither excessively expensive nor complicated to use, but to do your testing at the speed we are suggesting, you will need to test for continuity with the control circuit energized.

CHAPTER REVIEW

A primary goal of any maintenance electrician should be greater production time from plant equipment. Thus, the greater the electrician's speed in performing troubleshooting tasks, the less the downtime that will be accumulated on the equipment. The value of the maintenance electrician, then, is directly proportional to his or her troubleshooting speed and ability to get equipment back into operation.

A proper definition of electrical troubleshooting states that it is *the process of finding the first electrical or mechanical malfunction of any component part on which the normal operation of that electrical system is dependent*. In both troubleshooting and plant maintenance, industrial electrical equipment is viewed as a complete system.

Electrical safety must be uppermost in every electrician's mind. Whether the work being done is installation, maintenance, or troubleshooting, the most important safety consideration is the care and judgment used while working with electrical equipment. This is particularly true when working in close proximity to other circuits of a higher voltage. Poor judgment, sloppy working procedures, and high-risk shortcuts are an invitation to personal injury and equipment damage. Each set of working conditions will have its own safety criteria that extend beyond this chapter's generalized safety procedures.

THINKING THROUGH THE TEXT

1. In contrast to the practice of replacing entire systems because there has been a major shutdown, when should effective troubleshooting start? How will this enhance your employer's financial profitability?

2. Why is it important to be aware of the maintenance and troubleshooting of individual discrete electrical components in a system's circuit?

3. What is the difference between on-line and off-line testing?

    a. Give an example of on-line testing.

    b. Give an example of off-line testing.

    c. What is the primary advantage of on-line troubleshooting?

4. Identify the four Overvoltage Installation Categories issued by the International Electro-technical Commission (IEC). Why is the concern with the presence of transients important to the electrician's safety when using electrical test equipment?

5. What caution is given concerning multiple circuits in a panel?

    a. What different voltage values might be encountered in a single control panel?

    b. Which voltage values would most likely represent control circuits?

    c. Which voltage values would most likely represent working circuits?

CHAPTER

2

Understanding Electrical Symbols

OBJECTIVES

After completing this chapter, you should be able to:
- Identify the common electrical diagram symbols used in a ladder diagram.
- Understand the function of the conductors, switches, contacts, coils, and other devices represented by electrical symbols.
- Understand and be able to use ladder diagram number systems.

ELECTRICAL SYMBOLS

The focus of this book is electrical troubleshooting using ladder diagrams. However, before you can use a ladder diagram, you need to understand the individual **symbols** that depict the electrical equipment used in the circuit. You also need to understand the rules governing these symbols' structure and layout. Table 2–1 shows the electrical symbols you are most likely to encounter in industrial maintenance electrical work. Since you need to be familiar with all of these symbols and their functions, it will be helpful for you to stop and carefully look through the table.

As you study the symbols, you will realize that there is a simple logic in the graphic representations. Lines are drawn to represent current-carrying paths. Dots and circles are used to represent points where a conductor can be *touched* to make an electrical contact. Switches are drawn so that you can mentally *open* or *close* them, and so on. The following explanations should help you understand the information in Table 2–1.

Conductors

A conductor is represented as an unbroken line. Diagrams will show many conductors which are electrically common to each other. If there are only two electrical terminals that use this conductor, a single conductor line could be drawn between the two terminals. Figure 2–1 shows a single conductor as wire number 35 on line 19 because wire 35 has only two terminations; the first on relay cr10 and the second on relay cr3. If a single conductor is in contact with more than two wire terminals, additional conductors must be drawn. Figure 2–1 shows wire number 37 on line 19 as being common to two terminals on switch S3 and a terminal, which is not

14 CHAPTER 2 Understanding Electrical Symbols

Table 2-1. Common electrical symbols used in industrial ladder diagrams.

shown, to the further right of the diagram. Whenever wires are common on a diagram (electrically joined), they are shown connected by a heavy dot. Notice the dot on wire number 37.

On the other hand, because of the need to represent many parts of a circuit in a diagram, it is frequently necessary to carry a conductor line over another conductor line even though the two conductors are not common in the circuit. Consequently, the diagram must represent crossing lines that are not electrically common. When lines cross that are not common, a loop is used. Again, referring to Figure 2-1 you will see this loop drawn in the horizontal portion of wire number 18 on line 20 as it crosses wire 19. The loop indicates that the two wires are not electrically joined.

You may occasionally see diagrams that do not use loops on unconnected conductors. In this case,

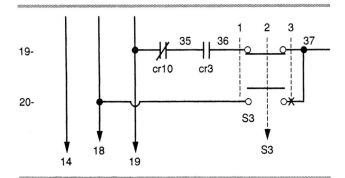

Figure 2-1 A heavy dot is the symbol used to represent a connected conductor. A loop is the symbol used to represent an unconnected conductor. Wire number 18 on line 20 contains both examples.

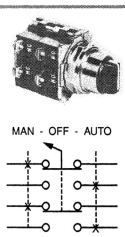

Figure 2-2 An industrial-rated selector switch and the symbol used to represent it. The symbol depicts a three-position, four-circuit switch. *Courtesy of Cutler-Hammer.*

you will see the heavy dot used for common (connected) conductors and the lines crossing but without the heavy dot for conductors that are not common.

On some diagrams you will see both heavy and light conductor lines. The heavy lines are used for high-current-carrying conductors in the panel, including the motor leads, the main disconnects and thermal overloads. The lighter lines are used for the lower-voltage control panel circuits such as switches, timers, and relays.

Switches

There are five common types of industrial switches used for machine controls:

1. Selector switches. **Selector switches** are **manual** rotary switches with two or more positions. Most are **maintained** switches, meaning that they will remain in any of their set positions. They are generally panel-mounted switches as shown in Figure 2-2. They are typically represented by a horizontal line in the electrical diagram as shown in the figure. The symbol in Figure 2-2 depicts a three position, four-circuit switch.

2. Push button switches. **Push button switches** are also panel-mounted switches. Though there are exceptions, typically they are **momentary**, two-position switches. A momentary switch must be manually held in its second position or it returns to its **normal state**. A push button switch is shown in Figure 2-3. Push buttons are represented by a horizontal line with a perpendicular center line as shown in the figure. The symbol in Figure 2-3 depicts a two circuit switch with a **normally closed (NC)** and a **normally open (NO)** contact.

3. Limit switches. Limit switches are generally machine-mounted switches that are used to signal the travel or position of moving equipment. Limit switches have numerous **actuator** designs to accommodate a variety of mechanical applications. Figure 2-4 shows a typical industrial limit switch. Limit switches are represented on the electrical diagram with a diagonal line, which

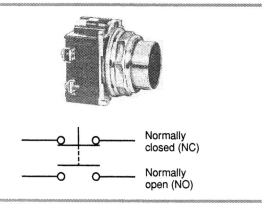

Figure 2-3 An industrial-rated push button switch and the symbol used to represent it. The symbol depicts a two circuit switch. *Courtesy of Cutler-Hammer.*

swings an arc to make electrical contact, and an attached ramp, which represents the mechanical switch function. Limit switches may have any one of four actions. In addition, they may be either held in position or maintained. *Held in position* means that this particular limit switch has a spring return that will restore the limit switch to its **normal position** unless it is otherwise prevented from doing so. A *maintained* limit switch does not have a spring return and will remain in its set position. These symbols are shown in Figure 2–4.

4. Process switches. Process switches signal changes in material level, pressure, temperature, or flow and are mounted on the machinery they are monitoring. Each category of **process switch** is represented by a unique electrical symbol as shown in Table 2–1.

5. Toggle switches. Toggle switches are less frequently encountered on industrial electrical diagrams. They are represented by a diagonal line, which swings an arc to make electrical contact. **Toggle switches** may have any combination of single line or multiple lines for separated circuits. They may have combinations of normally

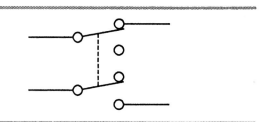

Figure 2–5 The diagram symbol for a toggle switch. This schematic shows a double-pole, double-throw switch used to create one normally closed circuit and one normally open circuit.

on and normally off circuits for the same throw. Figure 2–5 shows a double-pole (two-circuit), double-throw (two-position) toggle switch used to create one normally open and one normally closed circuit.

The electrical function of a switch in an electrical diagram is represented by two small circles on the termination point of two conductors. Making contact across the two circles creates an electrical bridge between the two conductors.

Within industrial electrical installations, the majority of selector switches and push button switches consist of the switch body, which is called an **operator**, and removable **contact blocks**, which make and break the electrical circuits.

The contact blocks in industrial switches are designated as being either normally open or normally closed. A normally open switch in its normal or relaxed state is nonconductive. A normally closed switch is conductive in its normal state; that is, in its nonoperated position. A diagram symbol identifies whether the switch is normally open or normally closed. If the switch is normally open, the conductive line is drawn above the two contact points so that an imaginary downward motion will make contact. If the switch is normally closed, the conductive line is drawn below the contact points so that an imaginary downward motion will break contact. Figure 2–3 shows both a normally open and normally closed push button symbol. A downward motion is generally the direction of activation for a hand switch or push button in electrical diagrams. It will return to the up position while at rest.

There is a technical qualification that must be added to the above paragraph. Selector switches use

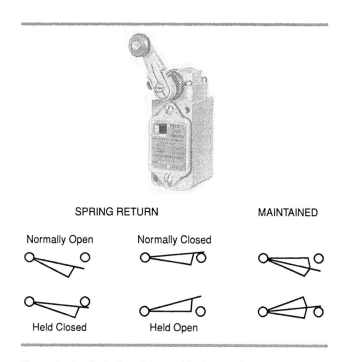

Figure 2–4 An industrial-rated limit switch and the symbol used to represent it. The symbols depict common configurations of a limit switch. *Courtesy of Cutler-Hammer.*

cams to press or release the protruding pins on their contact blocks. The particular cam of the selector switch will determine whether a contact block is in the held position or the relaxed position. For a simplified example, a manufacturer may supply a two-position selector switch that has a raised cam in the *on* (right) position and a lowered cam in the *off* (left) position. Using a normally open contact block would therefore give a made contact in the *on* position and an open contact in the *off* position. On the other hand, the same manufacturer may offer a second cam choice, which reverses the cam action and lowers the cam in the *on* position and raises the cam in the *off* position. You could use the second selector switch to accomplish exactly the same *off/on* operation by using a normally closed contact block. This has a practical application to our understanding of electrical diagrams. A diagram is not drawn with a normally open contact point because the equipment manufacturer used a normally open contact block. Rather, *a diagram is drawn with a normally open contact point because the switch and mechanical functions open the contact in what is understood to be the condition in which that contact would be in a relaxed state.* The same is true of normally closed contacts.

Figure 2–6 shows an exploded view of an industrial-rated push button. Most selector switches and push buttons are designed so that individual contact blocks are mounted to the switch body. In this way, various combinations of normally open and normally closed blocks may be chosen to accommodate the circuit design.

It will be helpful to understand how contact blocks are used to create individual switching circuits. With a two-position switch, a single normally open or normally closed contact block may be used to form each individual circuit in either of the two positions. For example, a two-position switch may be used to control two circuits where the first circuit is off/on and the second circuit is on/off. These two circuits could be simply controlled with a single contact block having both a normally open and a normally closed circuit. In the first position with the contacts relaxed, the normally open circuit would be off and the normally closed circuit would be on. When the selector switch is moved to the second position, the normally open contact would be on and the normally closed contact off. Each manufacturer will specify the maximum number of contact blocks that can be used on a single selector switch or push button. In the example shown in Figure 2–6, the manufacturer allows six contact blocks for a maximum of twelve individual circuits.

However, when three- and four-position selector switches are used, switching combinations must be created so that final circuit control is produced. This control sequence is created by coordinating the cam design of the switch itself with a combination of normally open and normally closed contact blocks in each switch position. To illustrate how this is achieved, look at the example in Table 2–2. In some cases, a particular switch combination may involve multiple contact block circuits. Table 2–2 shows four instances in which a circuit for a given position must

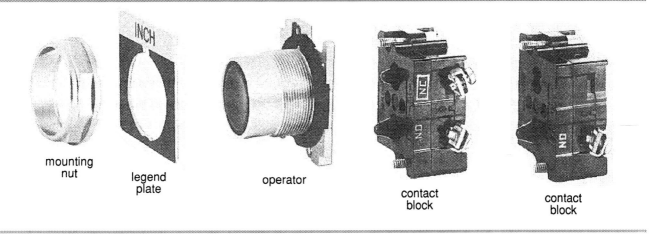

Figure 2-6 An exploded view of an industrial-rated push button switch operator and contact blocks. *Courtesy of Cutler-Hammer.*

Table 2-2 A contact block selection table for three-position selector switches. *Courtesy of cutler-Hammer.*

3 POS. SW — CAM AND CONTACT BLOCK SELECTION ❶

Combination No.	Desired Circuit Operation X = Circuit Closed O = Circuit Open ❷			Contact Blocks Required to Accomplish Circuit Function (Jumpers must be installed where indicated)			
				Operator with Cam Code #2 ❸		Operator with Cam Code #3 ❸	
				Mounting Location		Mounting Location	
				Top Plunger Ⓐ	Bottom Plunger Ⓑ	Top Plunger Ⓐ	Bottom Plunger Ⓑ
1	X	O	O	NO	NC	NO	
2	X	X	O		NC		NC
3	X	O	X	NO		NO	NO
4	O	O	X		NO		NO
5	O	X	X	NC	NO	NC	
6	O	X	O	NC		NC	NC

use two contact blocks. A similar table for a four-position selector switch shows some combinations that require three contact block circuits for a given position. However, you must be aware that the ladder diagram itself will represent the switch function as being a single contact.

Care must be taken while reading each electrical diagram. Hand and foot switches are generally drawn to indicate activation by a downward motion. However, most diagrams will show limit switches and process switches (level, pressure, temperature, or flow) as being activated with an upward motion. In each case, the electrician must study the individual electrical diagram to determine the orientation chosen by the draftsperson.

In the remainder of this chapter, you will be shown the relationship of the electrical symbols and the number or letter code systems with an actual ladder diagram of a working production machine. Figure 2–7 comes from the electrical diagram of a plastic injection molding machine. Throughout this book this machine's diagram will be used for many examples. As we did in Figure 2–7, we will often eliminate lines or details in order to conserve space.

Nonetheless, we will preserve the original line, wire, and component numbers in all examples. The complete diagram is given in Appendix A.

Switches and Contacts Are Shown in a Relaxed State

Most diagrams will show some switches as being *open* where the line representing the switch is not touching at least one of the circles, and some as being *closed* where the switch line is touching both circles. This is true of both normally open or normally closed switches. *The diagram always represents switches and contacts in the* off *or normal position.* For example, Figure 2–7 shows the motor start circuit. PBS (push button stop) on line 3 is closed until it is pushed. On the other hand, PBM (push button motor start) on the same line is open until it is pushed. Line 3 shows a door safety limit switch (LS10) that is open unless the door is latched shut. An *open* door is considered to be in the normal position. On line 6 there is a relay contact (cr1) controlled by relay CR1. The position of the relay contact is shown as if the power were off. In other words, with the power off, the relay contact between

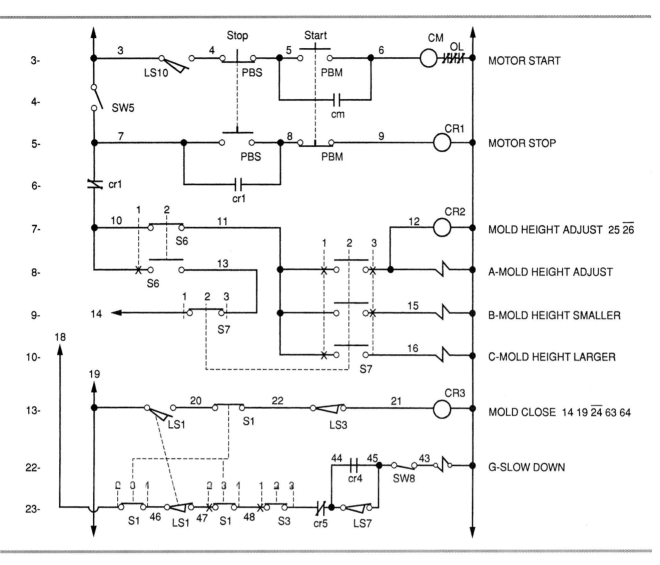

Figure 2-7 Selected diagram lines between 3 and 23.

wire 7 and wire 10 is closed. The slash indicates normally closed contacts. On the other hand, with the power still off, the relay contact between wire 7 and wire 8 is open.

The diagram also shows how a limit switch will actually function when it is cycled. Two limit switches are shown on line 13. Notice that the limit switch is always drawn so that there is a *ramp,* on the base of an inverted right triangle. You can visualize a roller moving along the ramp from left to right. If the roller is moved to the right, the switch contact arm on the two limit switches is moved up. This diagram assumes that the contact arm is controlled by gravity and wants to move to the down position. If the roller is moved to the right, the contact arm is raised and activated. Look carefully at limit switch LS1 on line 13 and you will notice that the contact arm is drawn below the contact circle on the end of wire 20. From that **schematic** representation, you can tell that LS1 (or that function because the limit switch may have more than one circuit) is an normally open switch that is closed when it is activated. On the other hand, limit switch LS3 on the same line is drawn showing the contact arm over the contact circle on the end of wire 21. Thus, you can tell that LS3 is a normally closed limit switch and will open when it is mechanically activated.

Limit Switches Have Multiple Functions

Limit switches have been designed to accommodate a number of signaling states depending on machine operation. For that reason, limit switches may

20 CHAPTER 2 Understanding Electrical Symbols

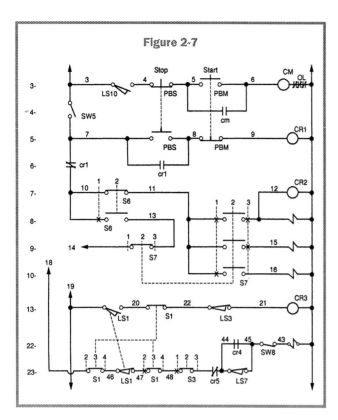

Figure 2-7

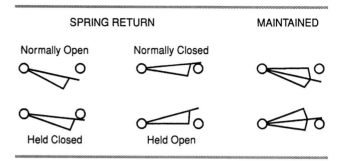

Figure 2-8 A normally open limit switch may be used in a machine as other normally open or held closed. A normally closed limit switch may also be held open. Limit switches may have maintained contacts where movement of the limit switch's actuator will change the switch state until the next cycle movement.

be either normally open or normally closed. However, these two configurations result in four possible schematic symbols. A normally open limit switch will rest in the open position. However, its use in a specific machine may require that in the normal (machine) position it be closed. In this case, the diagram would represent a normally open switch in the held closed position. The same possibilities are also true of a normally closed limit switch. A diagram may represent the normally closed limit switch as being either held open or closed depending on its normal state for that particular application. Limit switches may also be maintained. That is, a machine operation will alter the normal state of the limit switch to its second state on contact. The limit switch will maintain that state until a second machine operation returns it to a first state. Again, the normal state of the limit switch is shown as either normally open or normally closed. Figure 2–8 shows the applicable limit switch symbols.

Switches May Have Multiple Circuits

A single switch may have multiple circuits. Lines 8, 9, and 10 show four circuits contained in switch S7. The first circuit is common to wires 11 and 12. The second circuit is common to wires 14 and 13. The third circuit is common to wires 11 and 15. The final circuit is common to wires 11 and 16. A single switch may be used to control multiple circuits. Switch S7, for example, controls four circuits. In the case of switch S7, three circuits are common to wire 11. This is not always the case, however, because circuits controlled by a switch may be completely isolated. Figure 2–7 shows limit switch LS1 on lines 13 and 23. These two circuits are entirely separate. If you look at the complete diagram in Appendix A, you will see a third circuit for limit switch LS1 on line 43. Other types of switches may also have multiple circuits. Multiple circuits are often used in push button switches, selector switches, etc.

Do not confuse the terms *position* and *circuit* when referring to a switch or push button. You will be studying selector switch S7 in the next several paragraphs. You have just seen that switch S7 has four circuits. You will also see that it has three positions. The number of positions and the number of circuits are independent of each other. The number of positions is a mechanical function built into the switch. The number of circuits is determined by the electrical connections made through the contact blocks mounted to the switch operator.

If you understand the concept of a switch's multiple circuits, then you will understand the meaning of the connecting broken line. Again, look at Figure 2–7. You will see multiple switches or push buttons that are connected with broken lines. The vertical broken line between push button PBS on lines 3 and

5 indicates that the push button is mechanically connected so that pushing the button will move both contacting elements simultaneously, controlling each of the four circuits. The broken line indicates a mechanical function. *It is not an electrical conductor.* Do not read a broken line as a part of the electrical circuit.

Push Buttons

Push buttons are shown as an upside down T. The leg of the T is the thumb button, and the cross of the T is the electrical contact. If the electrical contact line is above the contact circles shown on the ends of the connecting wires, there will be a space between the circles and the line. This schematic representation indicates that the contact is open until it is pushed. If the electrical contact line is below the circles, the line will touch the circles. This second representation indicates that the push button is closed until it is pushed. Push buttons are always shown as being spring-loaded so that the button will return to an *up* (top-of-page) position. Look at the two push buttons on line 5 in Figure 2–9 for examples of a normally open and a normally closed push button.

There is a variation of what we have called an upside down T used to represent a push button. Table 2–1 also shows a mushroom head push button. When a push button is used as an emergency stop or panic button, it will often use a large head for both visual identification and ease of fast operation. This push button is identified by the curved line on top of the inverted T. However, the electrical function of this push button is unchanged.

For additional safety, an emergency push button switch may use a maintained contact. Table 2–1 shows the symbol for a maintained contact. In the case of an emergency stop, the maintained push button assembly will have a detent (without a spring return) so that the push button will remain in either the *down* or *up* position. For an emergency stop, this requires that the push button be reset by pulling the head of the push button to the *on* position. Maintained push buttons may frequently be used on equipment that must be in either one of two positions without an intermediate (or simultaneous) position. For example, a hydraulic cylinder control may use a maintained push button pair when a cylinder must be fully activated in a locked position or fully retracted in a loading position. In this case, the two push buttons will be mechanically linked so that pushing one push button releases the other. Both push buttons cannot simultaneously be in the activated position.

The opposite of a maintained contact push button is a momentary contact. The majority of push buttons used on a control panel will have a spring return so that the push button changes state only during the time the push button is manually held in the depressed position. When the push button is released, a spring will return the button head to its normal position. Most *start* and *stop* push buttons are momentary, spring return push buttons that depend on a relay to maintain the circuit. A *jog* function will also typically use a momentary push button. In this case, the motor being jogged will run only as long as finger pressure is maintained on the push button. Releasing the push button will stop the motor.

A push button switch, control switch, or limit switch may have multiple contacts. Rather than physically grouping all functions of a given button or switch together, the mechanical interconnection is represented by a broken line. Lines 3 and 5 in Figure 2–9

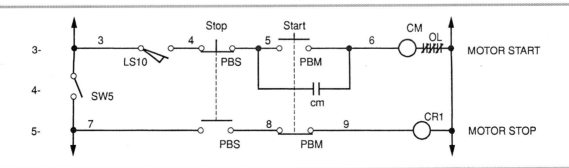

Figure 2-9 Diagram lines 3 through 5.

show two push buttons each. There are only two actual buttons rather than four, with each push button having two functions. The broken line between them shows that they are the same button. In the case of the push buttons on lines 3 and 5, the buttons are physically arranged on the diagram so that they are horizontally aligned. Alignment may not always be the case, as two functions of a single switch may be widely separated on the ladder diagram.

Notice in Figure 2–10 that even though the S7 selector switch functions on lines 8, 9, and 10 are physically close, the two functions on line 9 are not horizontally aligned, nor are they common to any of the same wire numbers. Separate switch functions are frequently encountered in limit switches, selector switches, and push buttons. Though the contact blocks are physically close to one another, the wire numbers and functions that they control may be widely separated on the diagram.

Selector Switch Functions

Selector switches are the most difficult to visualize because of their potential for multiple functions. Figure 2–10 shows switch S7 on line 9 between wires 14 and 13, which is diagrammatically connected (represented by the broken line) to the same switch on lines 8, 9, and 10. Selector switch S7 has three positions as indicated by the numbers 1, 2, and 3 above it. The broken line passes through the second position indicator. Thus, as indicated in the drawing, the contacts are either open or closed in the second position. For the other two positions (positions 1 and 3), the asterisk indicates when contacts are closed. On line 9, the single asterisk for position 3—adjacent to wire 15—indicates that wires 11 and 15 are closed when the switch is in position 3. However, they are open when the switch is in position 1 because there is no asterisk. You also know that it is open in position 2 because the drawing shows it as open. However, on line 8, the asterisks on both positions 1 and 3 indicate that the circuit is closed in positions 1 and 3 and the drawing indicates that it is open in position 2. Thus, the diagram shows which of the contacts are open or closed for each of the three positions. This switch also has four circuits. It controls a circuit between wires 13 and 14, and three circuits between wire 11 and wires 12, 15, and 16. Table 2–3 shows in table form the information, which is given by the asterisks and numbers on the ladder diagram for the three-position switch S7.

You will occasionally see another selector switch notation form that uses X to indicate closed, and O to indicate open. Thus, the switch function on line 8 would be indicated as XOX. The first switch function on line 9 between wires 14 and 13 would be indicated as OXO; the second switch function on line 9 would be indicated as OOX. The final S7 switch function on line 10 would be indicated as XOO. The simple two-position off/on switch on line 8 would be represented as XO.

Selector switches are available as both *maintained* and *momentary*. Most selector switches on industrial control panels are *maintained* switches, meaning they will remain in the position to which they are set. In some cases, however, a momentary function may be used. If, for example, a commercial printing press uses an ink feed pump that requires a *prime* function on initial start, a three-position, spring-return-from-right, maintained-in-left selector switch could be used. The switch would have three positions: *off, run,* and *prime*. The selector switch

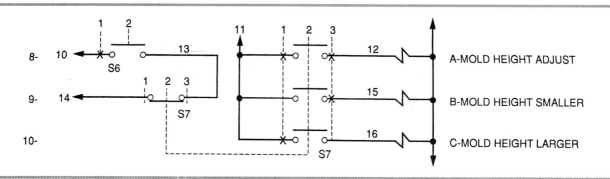

Figure 2-10 Diagram lines 8 through 10.

Table 2-3 Switch information represented by asterisks and numbers on a ladder diagram may also be depicted in table form. This switch table shows the electrical circuits generated by the three-position switch S7 in the diagram.

	Closed	Open
Position 1	11-12/11-16	13-14/11-15
Position 2	13-14	all others
Position 3	11-12/11-15	13-14/11-16

would remain in either the *off* or *run* positions as set. At start-up, however, the press operator could momentarily hold the switch in the *prime* position. In the momentary position, the pump would run while a prime solenoid was activated. When the selector switch is released, it will return to the *run* position and only the ink feed pump runs. The same functions could be duplicated with a two-position selector switch and a momentary push button. However, less panel space is required with the three-position selector switch as just described.

Other Switch Functions

Switch symbols include a number of specialized switches for sensing pressure, temperature, flow, etc. The need to know these symbols will primarily be dependent on your particular application. Do notice, however, the way in which they are schematically represented. The symbol designates whether they are normally open or normally closed.

Contacts and Timers

Contacts are generally divided into the two broad categories identified as **instantly operating contacts** and **timed contacts**. Instantly operating contacts (generally relay or timer contacts) are indicated by two parallel lines ─┤├─. Normally open contacts are drawn in this way. Normally closed contacts have a slash through them ─┤/├─. Timed contacts are generally those in which some sort of mechanical or electrical equipment delays the making (or breaking) of the contact after the initiation point. In early electrical panels, timed contacts were generally pneumatic dash pot units. These units used a moving diaphragm in an air-filled cylinder to delay the switching action. In newer electrical panels, you will see solid-state timers. These timers may be either fixed, having a preset time interval, or adjustable. As with all other symbols, contacts are indicated as though they are in the normal state.

Newer electronic timers are often supplied with many function choices. A single timer may be programmable for one of a number of timing functions. There are electronic timers on the market today with as many as six timing functions and two counter functions. They can be programmed from 0.1 second to 999 hours. These selections are all made with either thumb wheels or miniature switch selectors. Some of these single timer units can even be used on 24– to 240–volt alternating or direct current without any switching or modification. That is, the same timer can be unplugged from a 24–volt direct current supply and immediately plugged into a 240–volt **alternating current** supply with no adjustment to the timer. Needless to say, this greatly reduces the required timer/counter inventory in a manufacturing plant.

There are two timer functions represented in the electrical symbols that need to be explained. The first is identified as either **timed-closed** or **on delay**. *This timer begins timing when the signal circuit is applied.* After the timed interval, the relay contacts change state. In a practical application, this timer could be used to start a second motor with a single push button after a first motor had time to reach full speed.

The second timer is identified as either **timed-open** or **off delay**. *This timer begins timing when the signal circuit is removed.* After the timed interval, the relay contacts change state. In a practical application, this timer could be used to keep a conveyor belt running for a set period of time after the auger

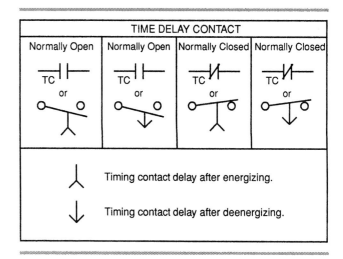

Figure 2-11 Symbols depicting timed-delay contacts.

24 CHAPTER 2 Understanding Electrical Symbols

feeding the conveyor is turned off. However, in both cases remember that the relays change state. Since there are both normally open and normally closed relays in the timer, the timer can be used to either energize or deenergize a control circuit (see Figure 2–11).

Coils

Figure 2–12 shows two symbols from Table 2–1 that need further clarification. These two symbols are always found on the right-hand side of the diagram. The symbol drawn as a circle is called a **shunt coil**. In most cases, you can think of the shunt coil as a low-**power component** (a relay, a timer, or a counter), which in turn is controlling a high-power component (such as a solenoid). The shunt coil may also be a panel light indicator. The sawtooth symbol is called a **series coil**. Relays, contactors, and starters are drawn as shunt coils. Solenoids are drawn as series coils.

Phantom Circuits and Supplementary Information

Referring to Figure 2–13, you will see a small circuit on lines 36 and 37 that has no physical connection with any other part of the circuit. This particular diagram is showing a **trigger circuit** for the counter, which is connected between the counter terminals 4 and 5. A trigger circuit carries no significant current but merely makes-and-breaks across two terminals of the counter or timer to initiate the

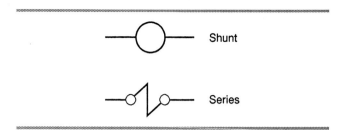

Figure 2-12 Symbols used for a shunt coil (top) and series coil (bottom).

counting or timing process. The diagram is simplified by showing only the counter functions rather than adding the complete counter schematic. The single function you need to identify is drawn with only the wire and terminal markings. The actual connections are not shown.

This circuit also shows wires 62 and 63 passing through a shielded conductor. The shield is represented by the cylinder symbol drawn around conductor 63. Because the counter contains electronic circuits, stray electromagnetic interference from the control panel's relays could produce inaccurate counts. This information on the drawing alerts the electrician to look for a shielded cable for this circuit rather than conventional panel wiring.

Because any given electrical diagram represents a specialized piece of equipment, there will be information on the drawing that is unique to that electrical circuit. This information is usually self-explanatory and can be understood by carefully studying the drawing. The information will often be

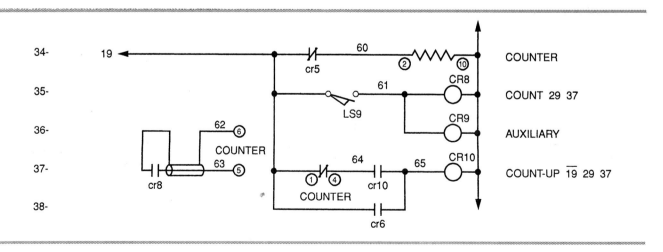

Figure 2-13 Diagram lines 34 through 38.

given as supplementary diagrams in the main drawing's margin.

DIAGRAM NUMBERING SYSTEMS

Most ladder diagrams will use numbers and letters to help you locate electrical components, numbered wires, and schematic locations. The following are the primary number or letter code systems you will need to use most electrical diagrams.

Line Numbers

On the left-hand side of the ladder diagram you will find a series of consecutive numbers from 1 at the top to whatever number of lines are included in the diagram on the bottom. The diagram you are using terminates with line 82 at the bottom as shown in Appendix A. These numbers merely identify the lines on the physical drawing. You will not find these numbers anywhere in the electrical panel. These numbers are only used as an aid in reading the diagram. The digits 3-, 4-, 5-, and 6- on the left of Figure 2–14 identify these four lines as numbers 3, 4, 5, and 6.

Wire Numbers

You will find a number (or letter) on the end of each wire throughout the entire machine. These numbers correspond to the wire identification numbers on the ladder diagram. Thus, on line 5 of Figure 2–14 you will see the three wire numbers 7, 8, and 9. If you were to open the control panel of the machine and look at either of the two push buttons or the relay shown on the diagram as being connected by wire 8, you would find wires with number 8 labels on their ends. The diagram shows these lines intersecting as a T. In fact, you will find four wire ends with number 8. A single wire end will be on any two of PBS (line 5), PBM (line 5), or relay contact cr1 (line 6). Two wire 8 ends will then be joined on a single terminal of the remaining connection point of PBS, PBM, or cr1. You would never expect to see a mid-wire splice on a control wire as assembled by the manufacturer. On the other hand, wire 9 (line 5) will only have two numbered ends. One end will be on push button PBM, and the other on CR1's relay coil.

Wires common to each other on the diagram are usually designated by a single number. For example, wire 7 (lines 4, 5, and 6 shown in Figure 2–14) has a single number designation on line 5. Yet, you know that all of the wires common to this numbered wire are also number 7 wires.

There is an exception to the number of wire ends you would expect to find. Wires from the machine will be bundled and brought to an electrical panel—usually in one or more conduits. Close to the entrance area in the panel, you will often find a long terminal block. The incoming wires will terminate on the terminal block with the control panel wires of the same wire number also running to the terminal block. Each junction point of the terminal block will have a wire number marker on it. Generally, the ladder diagram gives no indication of this terminal

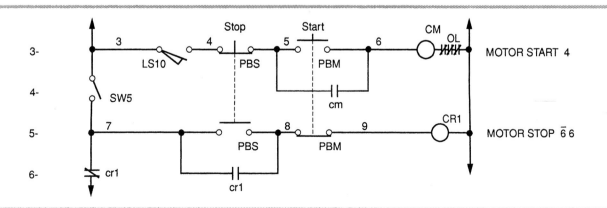

Figure 2–14 Diagram lines 3 through 6.

26 CHAPTER 2 Understanding Electrical Symbols

block, though you will find it to be an invaluable aid in troubleshooting.

Notice that on lines 70 and 71 of Figure 2–15 there is an arrow and a 46 (46 ←—) on the left portion of line 70 and an arrow and a 72 (72 ←—) on the left portion of line 71. This notation indicates an imaginary line drawn from the arrow to the wire number indicated. This separated circuit designation prevents complicated and confusing lines. It means that wire 46 from line 70 is connected to wire 46, which is found on line 23. Equally, on line 71, which is also shown in Figure 2–15, the arrow on wire 72 indicates that this wire is common to wire 72 found on lines 40 through 42. Notice that both line 72 and wire 72 occur on the same line. This is simple coincidence; you will occasionally see similarities between wire numbers and **line numbers** or wire numbers and component numbers. This will be the case when a line number contains a wire with an identical number. Similarities, however, are never intentional and have no special meaning.

Output Device and Component Letters and Numbers

Ladder diagrams will usually use both letters and numbers to designate **output devices** and power components. In the diagram you are using, relays are designated as CR1, CR2, and so forth, where CR is electrical shorthand for control relay. You will also notice that there are CR (cr) designations in both upper- and lowercase letters. The uppercase designation indicates the relay coil circuit. The lowercase letters indicate a set of contacts on that relay. Thus, CR1 in the diagram is the solenoid coil. There will also be a number of contacts on this same relay which will be designated as cr1 throughout the dia-

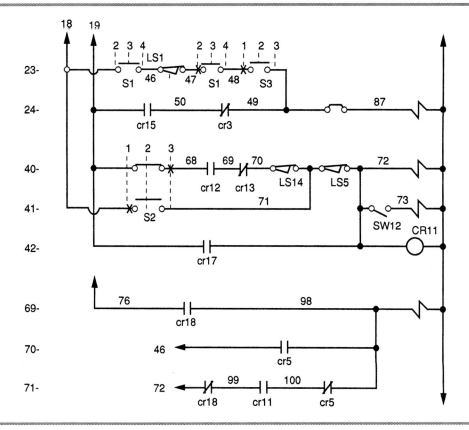

Figure 2-15 A diagram may use arrows to indicate common wire connections. On diagram lines 70, the arrow on wire 46 identifies this wire as common with wire 46 on line 23. On line 71, the arrow on wire 72 identifies this wire as being common to wire 72 on lines 40, 41, and 42.

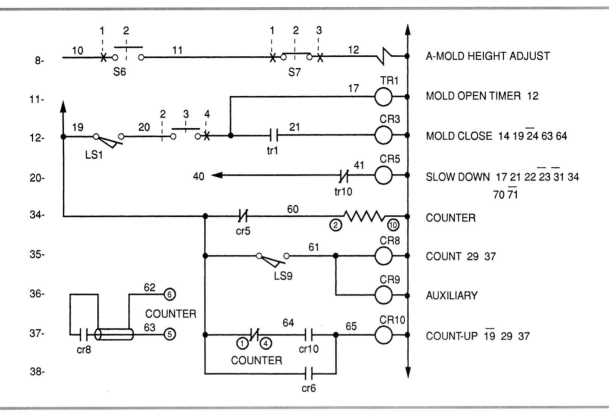

Figure 2-16 This ladder diagram contains number and letter information that identifies output devices and components. Their location on the diagram is also identified.

gram. Figure 2–16 shows CR8, CR9 and CR10 on the right-hand of the diagram. The relay contacts cr8 and cr10 are shown on line 37, while cr6 is shown on line 38.

In the paragraph above, the terms *output device* and *power component* were used. Because we will frequently be using these terms throughout the remainder of the book, a basic definition needs to be given for these and related words. When the words are being used in a technical sense, the following meanings apply:

Device. The *National Electrical Code®* defines an electrical "**device**" as *A unit of an electrical system that is intended to carry but not utilize electric energy* (Article 100). In this book, we will use *device* to mean any contact equipment that serves the purpose of opening or closing a control circuit. This will include all switches and push buttons because they *control* a circuit, but they do not *utilize* (consume) any electrical power. We will use two other terms identifying *device* when we need the precision.

Output device. The term *output device* designates a component that uses low power to perform the function of contact equipment. A timer, counter, or small relay all use some electrical power. Nonetheless, their function is to open and close a contact for the purpose of controlling other equipment. The output device is merely a means to the end of controlling a circuit.

Pilot device. The term **pilot device** indicates a control component that performs an identical function as the equipment which it controls, but is **interposing** between a low-power and high-power demand. Frequently in electrical control, a small contact in a miniature relay is used to control a power relay, which, in turn, controls a high-current load. Both relays function simultaneously. That which characterizes them, however, is the current they are capable of carrying. The distinction

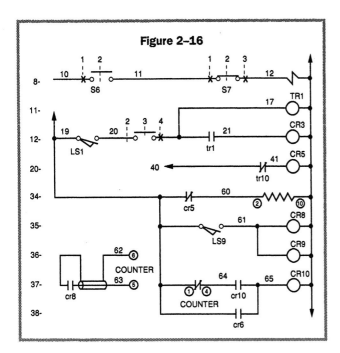

Figure 2-16

between an *output device* and a *pilot device* is merely one of **current-carrying capacity** and not electrical function. A motor starter is a pilot device, though we will generally use the familiar term *controller*.

Power component. Finally, the term *power component* or, merely, *component* identifies the end user of the electrical power. A motor, a heating element, and a solenoid are power components because they utilize electrical power to perform work.

Electrically operated hydraulic solenoids are designated on this diagram with letters as A, B, etc. Thus, A is the electrical solenoid pilot coil that activates the hydraulic circuit to make the mold height adjustment. The solenoid is shown on line 8. Timers are designated as TR1, TR2, etc. Again, as in the case of relays, there is a distinction made between the timer motor or solid-state timer circuit and the timer contacts. The timer motor is designated with upper-case letters as TR1, and the internal timer contacts with lower-case letters as tr1. Figure 2-16 shows TR1 on line 11 and tr1 on line 12.

Limit switches, selector switches, and the like, are generally identified in a legend portion of the diagram that describes their function. In Figure 2-16, limit switch 9 on line 35 is designated as LS9. On the legend section of the diagram, you will find a listing of all the limit switches, where you find LS9 designated as the "ejector stroke." Were you to go to the ejector hydraulic cylinder on the machine and physically locate this limit switch, there should actually be an identification label on the switch itself with the designation LS9.

Relay Contact Numbers

In some electrical diagrams, the actual relay contact used for each circuit is identified. On the diagram we are using, CR8 controls two circuits, CR9 is an auxiliary with no connections shown on the diagram, and CR5 controls eight circuits. CR8 could be a small relay with only two sets of contacts. CR10 could also be a small relay with only three sets of contacts. On the other hand, CR5 is a larger relay with eight or more contacts. If the diagram designated the wire positions, lines 34 through 38 would appear as shown in Figure 2-16. Each contact identification would show the wire position on the sequentially numbered relay.

Figure 2-16 also demonstrates another designation for relay, counter, and timer terminals. Since the terminals on these output devices have specialized functions, connecting line voltage to an inappropriate terminal may destroy a counter or timer. The counter is a **plug-in** unit that uses an 11-pin base. The terminal numbers are identified on both the counter itself and embossed on the base. Consequently, we know from the diagram that the counter's coil is energized from wire 60 on pin 2 and the common on pin 10. We also know that the counting signal comes from the second terminal on relay CR8 (designated as cr8 on line 37) to pins 6 and 5 on the counter's base. Likewise, we know that relay CR10 receives the count signal when wires 19 and 64 (line 37) change state from closed to open through the counter's normally closed relay. Counter pins 1 and 4 are the normally closed contacts.

Most relay configurations have a standard contact numbering system. The relay schematic and corresponding pin numbers are shown in Figure 2-17 for 8- and 14-pin plug-in relays. Most other relay configurations have standardized terminal numbering systems.

CHAPTER 2 Understanding Electrical Symbols 29

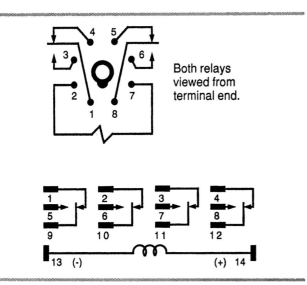

Figure 2-17 Plug-in relays have a standard pin or blade numbering system for the coil and contacts. The diagram views the terminal end of the relay; the base contact terminal numbers are the inverse.

Location Numbers

On the right side of the ladder diagram next to the common control wire is a series of descriptions and numbers. The descriptions tell us the function of that circuit by the output device it controls. For example, line 13 shown in Figure 2-18 controls relay CR3. Relay CR3 is identified as "mold close." To the right of the identifying description is a series of numbers. In this case, the numbers at the extreme right of line 13 are listed as 14, 19, $\overline{24}$, and 63. These are diagram line numbers, not wire numbers, where you will find contacts controlled by this relay. In other words, if you look on diagram line 14, you will find a set of contact points designated cr3. Similarly, lines 19, 24, and 63 will also have cr3 contact points. Notice the line drawn over 24 ($\overline{24}$). The overbar indicates that the contacts are normally closed. Look at cr3 on line 24. The points have a slash, indicating that they are, in fact, normally closed.

A PRACTICAL APPLICATION

One of the enjoyable aspects of learning is being able to see how new knowledge will help you in your daily work. Knowing the meaning of symbols on a ladder diagram will certainly help you find electrical problems as will be shown later in the book. But it may simplify other electrical maintenance tasks as well.

For example, you may need to repair a stalled machine after several wires were pulled loose from their terminals on a selector switch. In most cases, it

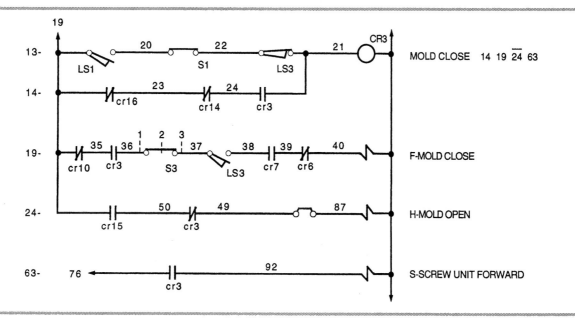

Figure 2-18 The location numbers for the mold close relay (CR3) indicate the diagram lines on which this relay is found (lines 14, 19, 24, and 63).

may be obvious which terminal is missing a wire. But this is not always the case. In some instances, a switch may have multiple circuits where two wires terminate on a single terminal.

Look at switch S3 on line numbers 15, 19, and 20 of Figure 2–19. What if the panel door had been jerked open and wires 14, 36, and 18 were all pulled from their terminals? Could you reconnect the wires to the correct terminals from the diagram? Reconnecting wire number 14 should be relatively easy. It would be reconnected on the terminal that was common to wire number 26 when the switch is in the number 2 position. But wires 36 and 18 are more difficult because they are both common to wire 37. However, the wires cannot be interchanged on the terminals because their sequence would be reversed. Interchanging wires could also create serious hazards. Your solution will come from the markings on the contact blocks, which will have either an NC or an NO printed on the block itself. After studying the diagram, you know that wire 36 must go to an NC block and wire 18 must go to an NO block. Before you energize the circuit, you could visually verify your work by putting the switch in the number 2 position and confirming that wire 36 is common to wire 37 while wire 18 is open. Then, by putting the switch in the number 3 position, you should be able to verify that 18 is common to 37 while wire 36 is open.

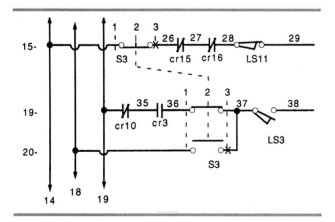

Figure 2-19 The electrical diagram identifies wire locations. This portion of the diagram can be used to reconnect wire numbers 14, 18, 26 and 37 on selector switch S3.

CHAPTER REVIEW

Electrical symbols and diagram formats have been standardized in an attempt to simplify reading electrical prints. With little exception, if you are familiar with the symbols in Table 2–1, you will be able to read almost any American industrial electrical diagram.

Electrical symbols are a graphic representation of electrical circuit logic. Conductors and conducting elements in switches, relays, and other electrical components are represented by lines. Controlled parts of the circuit such as relays, switches, push buttons and timers are represented by graphic symbols that will complete an open part of the circuit. The diagram always represents switches and contacts in the off or normal position.

The complete ladder diagram uses number and letter code systems to designate specialized information. Line numbers on the extreme left of the diagram are solely for the purpose of diagram referencing. Wire numbers are markings on both the diagram and the wire itself, which identify individual conductors. Each common conductor has a unique wire number. Power component and output device numbers are used to identify the physical parts of the electric circuit. Upper-case letters (with numbers) identify the electrically activated coil or motor of the output device. Lower-case letters (with numbers) indicate the controlled contacts of that same output device. Other number codes include terminal numbers, which are used to indicate the appropriate terminal location for a given wire, and location numbers on the right side of the ladder diagram, which are used to indicate the location of the contacts controlled by that output device.

THINKING THROUGH THE TEXT

1. Reproduce the following symbols shown in Table 2–2 and describe their electrical function:

   a. A normally open limit switch

   b. A normally open, held closed, limit switch

c. A normally closed liquid level switch

d. A normally open temperature activated switch

e. A normally closed flow switch

f. A normally open push button

g. A normally closed push button

h. A maintained push button

i. A normally open relay contact

j. A normally closed relay contact

k. A connected conductor

l. A non-connected conductor

m. A shunt output coil

n. A series output coil

2. Describe the meaning of a wire number.

3. Describe the meaning of a line number.

CHAPTER 3

Understanding Ladder Diagrams

OBJECTIVES

After completing this chapter, you should be able to:
- Understand the relationship between electrical symbols and the ladder diagram.
- Identify and use either a ladder diagram or a wiring diagram for electrical troubleshooting.
- Understand an electrical system by reading its ladder diagram or wiring diagram.
- Visualize the function of individual electrical components within an electrical circuit.

INTRODUCING THE LADDER DIAGRAM

The electrical diagram most frequently used for large and complex circuits in plant maintenance work is called either a ladder diagram or a line diagram. The ladder diagram is a schematic representation of the electrical circuit. It is drawn in an H format, with the energized power conductors represented by vertical lines and the individual circuits represented by horizontal lines. The electrical diagrams used in Chapter 2 showing circuits from a plastic injection molding machine were all in ladder diagram format. Refer to Appendix A for a complete ladder diagram. Notice that the ladder diagram is defined as a *schematic* representation of the circuit. It is not a *physical* representation. The electrical components and conductors are schematically arranged according to their electrical function in the circuit. The wiring diagram—which will be discussed briefly at the end of the chapter—is a physical representation. In the wiring diagram, the electrical components and conductors are arranged on the basis of their actual physical relationship to each other.

The Ladder Diagram Simplifies Circuit Representation

Simplicity is the purpose of the schematic layout of the ladder diagram. Diagram complexity is greatly reduced by indicating each circuit as a single horizontal line. Compare the difference in complexity between the wiring diagram in Figure 3–1 and the ladder diagram in Figure 3–2. Both represent the same compressor motor circuit. Notice the larger number of lines and complexity in tracing the circuit in Figure 3–1.

CHAPTER 3 Understanding Ladder Diagrams   33

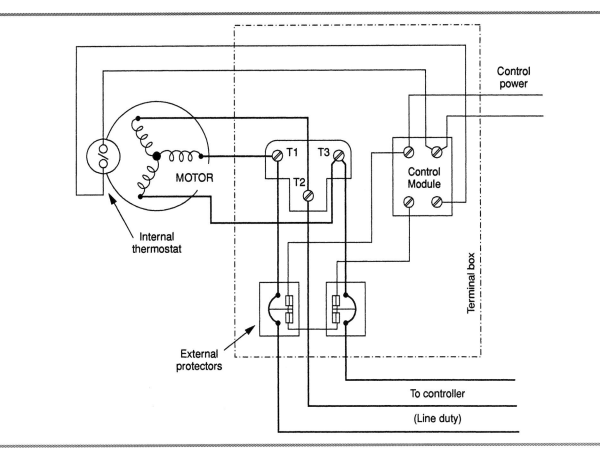

Figure 3-1 A wire diagram for a refrigeration compressor motor. This is a physical representation of the circuit and requires tracking conductor lines to separate locations on the drawing.

Because the ladder diagram is not concerned with the physical relationship of the machine circuits, it can be used to represent very large electrical cir-

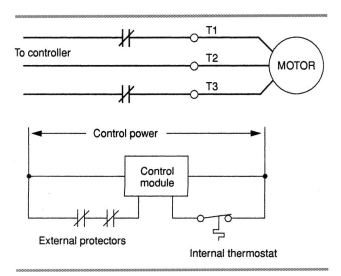

Figure 3-2 The same refrigeration compressor motor circuit drawn as a ladder diagram. The ladder diagram is logically more simple but it does not physically represent the equipment.

cuits by merely increasing the length of the circuit diagram. This does not add to the difficulty of reading individual circuit lines. As a matter of fact, when you fully understand the ladder diagram, you will appreciate its genius for reducing a complex electrical circuit to a manageable graphic representation.

The Ladder Diagram Is a Logical Circuit Representation

Using the ladder diagram will be relatively simple now that you understand the meaning of the electrical symbols. The ladder diagram merely represents the current paths shown as **rungs** of the ladder leading to each of the controlled or energized output devices or power components. As you read from left to right along a ladder diagram line, you will see each of the electrical devices that control the electrical current to a specific output device or power component.

Figure 3–3 shows the motor-starting circuit from diagram *lines 3 through 5*. This is the current path needed to energize and lock the motor's magnetic

34 CHAPTER 3 Understanding Ladder Diagrams

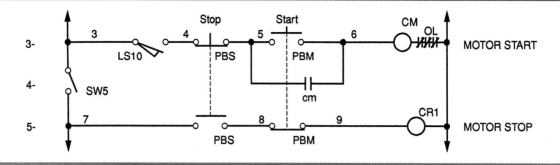

Figure 3-3 Diagram lines 3 through 5.

contactor, which is designated as CM. For the motor contactor (CM) to cycle, wire 3 must be energized and a series of electrical contacts must close. The ladder diagram shows each of the contacts through which the circuit for a given electrical output device or power component must pass. As you read from left to right along line 3, you see that the circuit will pass through four individual contacts: the rear gate safety (LS10), the motor stop push button (PBS), the motor start push button (PBM), and the motor overload protection circuit (OL) on the right side of CM. In the machine itself, the electrical current will follow exactly the same path as shown on line 3 to reach the motor contactor coil (CM). It is the purpose of the ladder diagram to graphically represent the exact current path through each electrical device in the electrical panel.

You should be able to immediately see the value of the ladder diagram as a troubleshooting aid. As you will see later in the troubleshooting section, this diagram will allow you to follow the path of the circuit until you find the one electrical device or component that is failing to function. Since the ladder diagram is an exact representation of that current path, you can do various voltage or continuity checks along the path to verify the operation (or failure) of each device or wire connection.

The Ladder Diagram Represents Only Electrical Functions

You must remember, however, that the ladder diagram represents only the equipment's electrical functions. It does not represent the physical wiring layout of the machine. You will see sections of the panel wiring where the physical locations of various components are adjacent to each other and are connected by a single wire just as it is indicated in the diagram. However, the wiring layout in terms of the physical location of components is often quite different from the ladder diagram representation. This is true simply because switches and push buttons are often in quite widely separated locations on the actual machine.

The number of lines used on the drawing are kept to a minimum. An example of the reduction of lines is the indication of a common point by a heavy dot and intersecting lines. For an example of a midwire connection on the diagram, look at wire 5 on lines 3 and 4 in Figure 3-3. However, as you saw in Chapter 2, this is not how you will find the wiring physically arranged in the machine. Wires usually terminate on electrical device or component terminals or junction block lugs; mid-wire splices are rare. Unless the terminals are specifically rated for three or more wires, you would find no more than two wires under a screw terminal. You would also expect to see electrical components widely separated on the actual machine, although they are represented on the diagram as being adjacent to each other. Limit switch LS10 and the stop push button PBS are adjacent to each other on line 3. However, their location is widely separated on the machine. The start push button is on the operator's panel while the limit switch is the door safety on the opposite side of the machine from the operator's panel. With these basic qualifications concerning the physical arrangement of the circuit, however, you will find the ladder diagram to be a great help in electrical troubleshooting.

CHAPTER 3 Understanding Ladder Diagrams

READING THE LADDER DIAGRAM

For you to effectively use the ladder diagram, there are some relatively simple details that you need to understand concerning the physical representation of the diagram symbols.

Vertical Diagram Lines

On the right side of the diagram, there is almost always a common **grounded conductor** represented by a vertical line. In the diagrams we are using, this is the vertical line on the right side of the ladder diagram. If you look at the complete diagram in Appendix A, you will see that this wire runs to the bottom of the diagram. The common grounded conductor will be identified by either a number used on both the diagram and on all of the appropriate wire ends, or a unique wire color in the machine. Some diagrams will use L1 for the incoming energized line and L2 for the grounded conductor. The diagram you are using does not use a number for the common grounded conductor. However, in the electrical panel, this wire is always white. All other wires in this same panel are red with a number band on each end.

There are exceptions to the statement that the right-hand side conductor is always grounded. If the control system is 240 volts AC with a 120-volt center tap, both high voltage lines will be fused and ungrounded. Though well-grounded systems are generally used with Programmable Logic Controller (PLC) equipment, you may encounter a PLC installation in which there is no grounded common conductor. Older installations, equipment built on sight, or imported equipment may not have proper grounding of the neutral conductor. If the neutral is not properly grounded on the equipment you are servicing, it should be brought to *NEC®* standards whenever possible. However, if the installation includes PLCs or other electronic equipment, the change should be discussed with the manufacturer before proceeding.

The common grounded conductor is also shown as the direct electrical connection to one side of the output device or power component that is being controlled on that line. This output device or component uses current to do work. It may be a relay coil, a hydraulic solenoid, a timer motor, or even a buzzer.

The left side of the ladder is also represented by vertical lines, although these lines are not necessarily common to each other. However, significant sections of the diagram often are common.

Figure 3-4 identifies a number of common design features of a typical 120-volt panel control

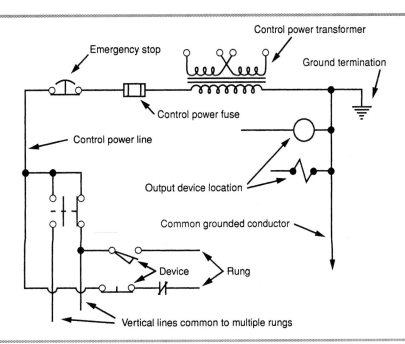

Figure 3-4 The ladder diagram represents the power source (transformer) at the top of the diagram. The fuse protected common power circuits are represented with vertical lines on the left. The common grounded conductor is represented with a vertical line on the right. The output devices are always on the right adjacent to the grounded conductor.

system. Notice that the fuse is directly adjacent to the **transformer secondary** with an emergency stop immediately following. The common grounded conductor terminates on the unfused side of the control transformer and is common to the ground jumper. In practice, the ground jumper is an individual wire run directly from the transformer terminal to the chassis ground; the wire is as short as practical, usually not over 10 inches long. Figure 3–4 shows the output devices and power components immediately adjacent to the common grounded conductor. The figure also indicates that control power lines on the left-hand side may be common to many rungs of the control circuit. The lines on the left-hand side of the ladder diagram will be connected to a terminal on a switch or other controlled current source.

Ladder Diagram Contact Positions

An important observation should be made here concerning the reading of the ladder diagram while doing troubleshooting. Remember, the ladder diagram is drawn as though the control voltage is *off*. In other words, as though the circuit is dead. When activated either electrically or mechanically, the contacts will be in the reverse position from that shown on the diagram. Look at lines 3 and 4 in Figure 3–5. The push button is pressed to start the motor that closes PBM between wires 5 and 6. PBS must also be closed between wires 4 and 5. Since the stop function is a normally closed push button, that part of the circuit will be complete unless the button is pressed, or unless there is a malfunction in PBS. LS10 must also be closed before the motor will start. This means that the rear door on the machine must be closed. If the circuit is complete, the motor will start when the motor controller CM is energized. However, if you remove your finger from PBM, you will open the circuit and the motor will stop. So a **holding**, or **latching**, **circuit** is provided on line 4. When CM is energized, it closes contact cm between wires 5 and 6. This keeps the motor running until it is stopped by pressing PBS or opening the rear door (LS10).

Notice that the holding circuit (relay contact cm on line 4) is a contact on relay CM itself. This is why this circuit is called a *holding* circuit. Once relay CM is energized, it will remain energized until power is interrupted on wire 6. Notice, that the stop push button (PBS) and the rear door safety limit switch (LS10) are both placed *before* the holding contact between wires 5 and 6.

There is also motor overload protection on the circuit. Between the motor contactor (CM) and the common grounded conductor, there are three sets of contacts in series that are shown in a normally closed position. There is one contact monitoring each phase of the incoming motor circuit. If any one of the three incoming lines draws excessive current, the overload will release and stop the motor.

If you are checking the circuit of Figure 3–5 on the machine because the motor will not start—or will not continue to run—you will be testing a circuit that ideally should have LS10 closed rather than open as shown in the diagram. PBS should be closed as it is shown in the diagram, PBM should be closed when starting and cm closed when running, although both are shown as being open in the diagram, and the overloads (OL) should be closed as shown in the diagram. *When troubleshooting, a great deal of confusion will result if you do not keep in mind that some contacts will be in the opposite position when energized.* Further, you will need to be able to visualize whether a contact is to be read as energized or de-energized for the condition or setting of the machine during the actual test. Look again at line 3 in

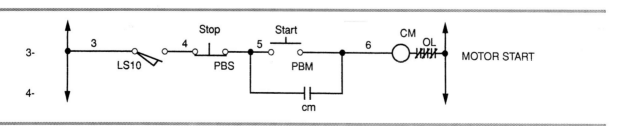

Figure 3-5 Diagram lines 3 and 4.

Figure 3–5. When the motor is running—or the circuit is energized—you cannot merely read everything as being reversed from the diagram representation. Some symbols will be reversed, while others will not. For example, in the normal running condition, the thermal protection for the motor (OL) will not be open. It will be in the conductive state as shown in the diagram. Equally, PBS will be in the normal position and, therefore, will be in the conductive state as shown in the diagram. On the other hand, if the circuit has been energized and the motor is running, the rear gate safety switch (LS10) and the motor holding circuit contact (cm) will be energized, or closed. Thus, when you are reading a diagram during actual testing, you need to be aware of each symbol's function and decide whether it should be read as shown in the diagram, or read in its energized or activated mode.

Horizontal Diagram Lines

The primary function of the ladder diagram is its line representation of the current path to each output device or power component. Since an electrical device will control an end function either directly through a relay, or indirectly through a pilot device or power component such as a hydraulic control valve, a failure of that end function to activate will indicate which part of the circuit has malfunctioned. This is assuming, of course, that the failure was caused by the electrical control circuit. If you know which part of the circuit has malfunctioned, you can refer to the lines in question on the ladder diagram and begin your troubleshooting procedures. You will be looking for the failure of specific switches and contacts on that line of the diagram. However, you will see later that a contact not closing on the line you are checking may be caused by a malfunction on another line. Often, a relay contact is not closing on your first line, but the cause of the problem is actually on the line that is controlling the malfunctioning relay's coil. For example, refer to the ejector section of the diagram in Figure 3–6. If the ejector does not move forward, you know that one or both of the hydraulic pilot valves (L "ejector" or M "ejector pressure" on lines 31 or 32) are not activating. Since the two solenoids are controlled by a **parallel circuit** with no other relay contacts or switches between them, you know that the problem is on lines 29, 30, 31, 32, or 33. You would then need to check only the devices on those lines or the wires feeding those lines, which would be wires 14 and 19.

The actual troubleshooting procedure will be explained in Chapter 7. However, from the example you have just been given, you can see how quickly an area of the circuit can be isolated for further troubleshooting. It was isolated simply by observing the end function, which did not activate, and then going quickly to that section on the ladder diagram. If you do not believe that what you have just been

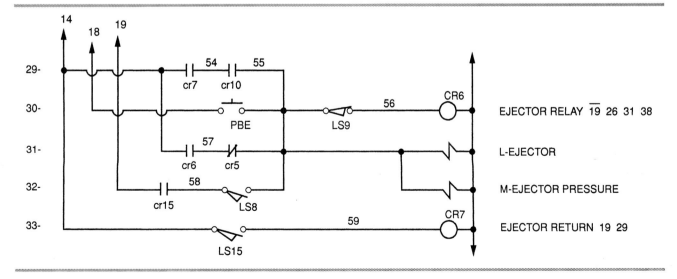

Figure 3-6 Diagram lines 29 through 33.

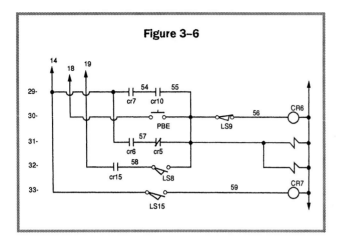

Figure 3-6

told is an example of "fast," then you need to spend a day watching an "expert" who does not need diagrams tear the innards out of both the hydraulic system and the electrical panel only to find that there was a loose wire terminal on wire 57 to cr5 on line 31!

UNDERSTANDING THE LADDER DIAGRAM

In Chapter 1, you were told that good troubleshooting was based on understanding the electrical circuit as a system rather than as a series of discrete electrical components. This should become obvious to you as you study the complete ladder diagram.

The Interrelated Circuit Diagram

In the last several paragraphs, we introduced an important troubleshooting concept. The interrelatedness of the electrical system is most evident as you follow a circuit represented by one of the numbered lines. Though it is not always the case, you will frequently find that a given line will be dependent on electrical output devices in other lines. In order for the relay on the line you are tracing to function, its circuit is dependent on other relays, timers, and the like, which must also operate normally. Look at Figure 3–6 again. You will see that CR6 depends on four other relays in order to function. Relay contacts are shown for CR7 (cr7 on line 29), CR10 (cr10 on line 29), CR5 (cr5 on line 31), and CR15 (cr15 on line 32). In addition, on line 31 there is a relay contact in CR6 itself (cr6) that controls its own function. Not all five of these relays are in the circuit at a given time, however. Some are only used in the **automatic mode**, while others are used in only the manual mode. If you were troubleshooting this circuit because CR6 was not functioning properly, you would possibly need to troubleshoot the diagram line for any one of the other four relays. As an example, a damaged relay coil on CR7 would cause relay CR6 to fail in the automatic mode.

Thus, as you learn to read ladder diagrams—and when you are actually doing the troubleshooting work in the electrical panel itself—you will need to be aware of the relationship of the line you are working on to other parts of the circuit on which it is dependent. Though it is not always the case, relay contacts on the line you are checking are frequently controlled by relays and switches on other lines. You will need to work with multiple lines in your testing, though once you understand on-line troubleshooting, it should not be too difficult.

Proper Circuit Grounding

Before leaving the subject of ladder diagrams, there is an area of technical information that you should understand. The electrical output device or power component is always located at the extreme right-hand end of the circuit, next to the common grounded conductor. There is a very important safety reason for this location. Figure 3–7 shows a ground symbol on the grounded neutral wire. When the circuit is energized, the common wire is grounded to the panel. That means that the entire circuit to the left of the electrical output device or component is the ungrounded circuit. It is also called the "hot" leg. Therefore, *any short to ground in the entire circuit will blow the control transformer fuse.* The fuse is shown on the left-hand side of the diagram immediately adjacent to the control transformer secondary.

What would happen if the output device or power component was moved from the right side of the diagram to the left side? For the sake of this example, the left side of the diagram will still be the "hot" leg. A short to ground any place in the control circuit would *activate* the output device or component because the control panel chassis would complete the circuit. Can you imagine the danger that would result? A **short-circuit** to ground would not blow the fuse. *It would activate the circuit.* A hydraulic cylinder could slam shut because of a ground fault! Equipment damage and personnel injuries would be very likely. Of course, switches and contacts between the output device or component and the short must still be closed. However, looking at the bottom rung on Figure 3–7 will show you that a short to ground could still bypass the limit switch safety. At the very least, a short could cause a function to cycle out of sequence with equally disastrous results. You can see, then, why the output device or component is immediately adjacent to the grounded side of the circuit. Anything added to the right of the output device or component would cause it to function in the presence of a short to ground rather than blowing the fuse. The only exception to this rule is the motor contactor overloads, which are placed between the coil and the grounded neutral. If you ever do any add-on wiring in a panel, *always* add it on the hot side of the circuit.

Figure 3–7 also indicates that there is only one fuse on the circuit side of the control transformer. There is a fuse on the "hot" side, but not on the grounded neutral side. Again, there is an important safety factor in this design. If a fuse was placed on the grounded side and it opened—but the fuse on the hot side remained conductive—the circuit would be inoperative, but the conductors would remain live.

The circuit could not function, but it would present a hazard to anyone working on it. In practice, a fuse is never placed on the grounded side of the circuit. *NEC® 240-22* states that, *No **overcurrent** device shall be connected in series with any conductor that is intentionally grounded.*

LINE CONDUCTOR DIAGRAM

In this book the primary emphasis is troubleshooting with the low-voltage control circuit diagram. The majority of your industrial plant troubleshooting time will be spent with control panel problems.

However, you also need to understand the line conductor diagram, which shows the high-voltage circuits to the motors and transformers. You will occasionally need this diagram for troubleshooting. You will most certainly use it when you are installing new equipment.

Typically, this diagram will show the main power conductors with heavy lines. Notice how Figure 3–8 uses heavy lines between the main disconnect and all primary fuses. The line conductor diagram will include information regarding the particular piece of equipment you are servicing. The diagram in Figure 3–8 includes motor horsepower and motor connections for 240– and 480–volt installations. Notice that

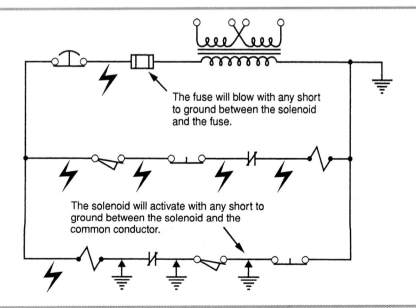

Figure 3-7 When the output device is properly located next to the common grounded conductor, a short-circuit to ground will blow the fuse. If an output device is improperly located with control devices between it and the common grounded conductor, short to ground can activate the circuit rather than blow a fuse.

40 CHAPTER 3 Understanding Ladder Diagrams

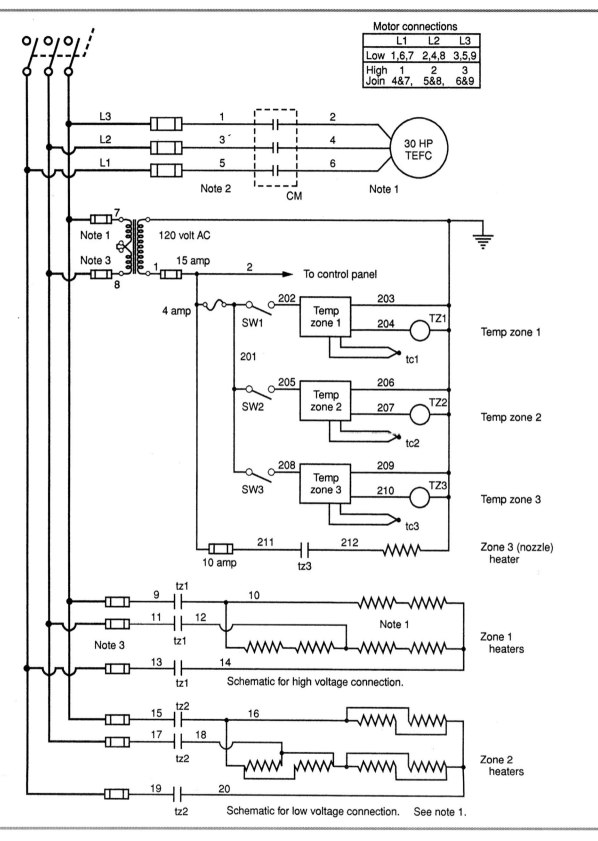

Figure 3-8 The line conductor diagram represents the high-voltage part of the electrical system. This diagram may also give useful information concerning voltage selection and installation data for the machine.

this diagram also shows the appropriate heater connections for either 240–volt (parallel) or 480–volt (series) installations.

The injection molding machine we are using as our example has a single motor for the hydraulic pump. Most line conductor diagrams for larger equipment would indicate a number of motors and their individual overload protection circuits. However, this diagram includes the heater controllers and heater elements for the machine.

WIRING DIAGRAMS

Wiring diagrams will not be used for troubleshooting examples in this book. Nonetheless, they are a useful source of information that you will use in your electrical work.

The wiring diagram's primary role is that of representing a specific—and usually limited—circuit function. In Figure 3–9, a refrigeration compressor motor circuit is shown. The purpose of this diagram is to identify the electrical functions and physical arrangement of one circuit area. To that end, the wiring diagram is the best means of representation.

You will often find these diagrams in cover plates and control boxes. Frequently, there will be a wiring diagram inside a motor contactor cover. Because these circuits are small, it is advantageous to use physical rather than schematic diagrams. It is simpler to identify terminals and wire locations from a physical diagram. Occasionally, for equipment with simple circuits, you may find that the entire unit is represented with only a wiring diagram.

Frequently, the wiring diagram will include other useful information. Again, the wiring diagram inside a motor contactor cover will list information pertinent to that model starter such as its horsepower rating, the overload coils to use for various current loads, and nonelectrical information such as the replacement part numbers for that particular unit. The wiring diagram inside a motor cover plate may list fuse sizes or overload values relative to that particular motor. If you can decipher the fine print, there is often information on these printed diagrams that

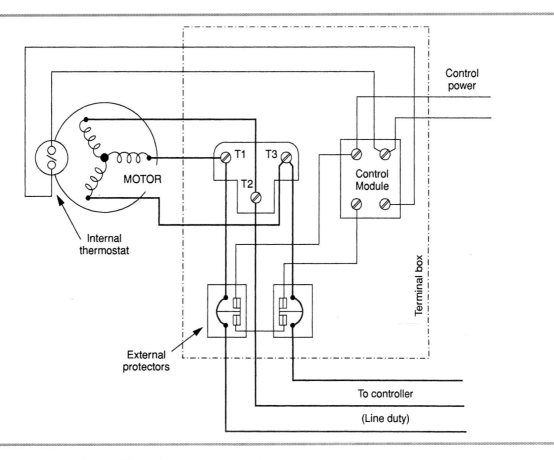

Figure 3-9 A wire diagram for a refrigeration compressor motor.

may be helpful to you in your troubleshooting work. The solution to a motor tripping the overloads may be as simple as comparing the horsepower ratings with the overload chart in the cover plate. Pay particular attention to the allowable ambient temperature ratings for a given size heater.

When using a wiring diagram, you will read it in much the same way as you would a ladder diagram. The circuit functions are represented by their appropriate symbols. However, you will need to follow conductor lines throughout the drawing rather than a single horizontal circuit line. This will add to the potential confusion of this type of diagram. The wiring diagram is a useful circuit representation. Its use, however, is generally limited to small, specific circuit functions.

Specialized Wiring Diagrams

Occasionally wiring diagrams are used to represent more complex circuits. Figure 3–10 is the wiring diagram from a wire feed machine for a welder. This diagram shows the physical layout. Notice that the wiring of the plug connectors to circuit boards is exactly as it is found in the machine itself.

The great difficulty in using this type of wiring diagram comes in tracing a given wire number through the schematic. As an example, pick various wire numbers in Figure 3–10 and trace them through to their termination points. Notice that you must pay close attention to the direction of the branch sweeps to determine if a wire could, or could not, follow a

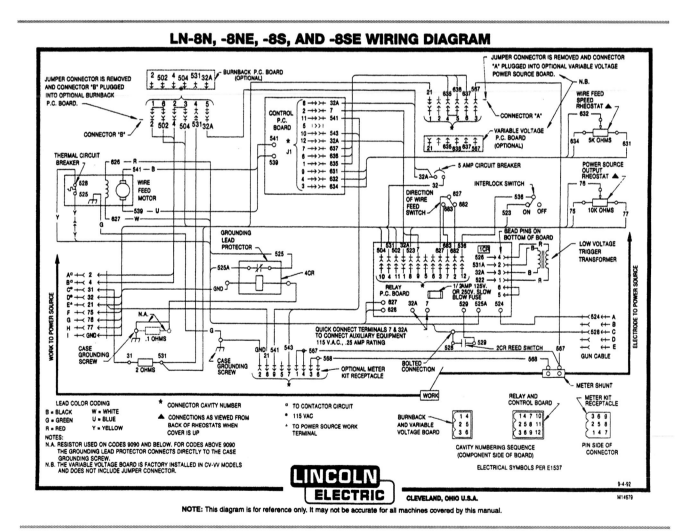

Figure 3-10 A more complete wire diagram for a welder wire feed machine showing wire number location on individual circuit boards, components, and connectors. *Courtesy of Lincoln Electric Co.*

given path. Notice also that some wire branches do not terminate in a single location.

Though the diagram used in Figure 3–10 shows few components, the same format may be used to show a full complement of electrical equipment. In this latter case, the wiring diagram is much more difficult to use than a ladder diagram.

On the other hand, there are reasons why this type of diagram is used. More non-circuit information can be put in a smaller drawing space when these diagrams are placed inside machine panels. In addition to the electrical schematic, these diagrams will also include receptacle pin information, jumper wire terminations, and other details, which are better communicated with a physical diagram layout.

CHAPTER REVIEW

The ladder diagram represents the individual current paths for each of the controlled or energized electrical output devices or components. The current path is represented as a conductor line containing symbol representations of each of the electrical devices that are found in that particular circuit. The ladder diagram is a representation of the electrical function of the machine, though it is not a representation of the actual physical layout.

The ladder diagram is always drawn in the deenergized state. Mechanically operated devices are represented in the open or normal position. When the diagram is being read in actual testing with an energized circuit, symbols must be appropriately read as being either energized or deenergized, depending on the state of that particular circuit.

Both the ladder diagram and the actual circuit always connect one leg of the electrical output devices or components directly to the common grounded lead. This connection arrangement will blow the fuse or **circuit-breaker** in the event of a short-circuit to ground anywhere in the protected circuit. This is a mandatory safety precaution that prevents a short to ground from activating the output device or component.

The grounded common wire from the control transformer is never fused. Blowing a fuse on the grounded leg would deactivate the circuit, but it would create a hazard by leaving the ungrounded side of the circuit energized.

The wiring diagram is generally used to represent limited circuit functions. It is a physical representation of the circuit, showing the various electrical components as they are related to each other and the conductors for that particular circuit. Wiring diagrams are often found inside terminal and panel covers. They often contain specific electrical and nonelectrical information for that particular unit.

THINKING THROUGH THE TEXT

1. What is the value of the ladder diagram as a troubleshooting aid?

2. The ladder diagram represents the equipment's electrical functions but not the physical wiring layout of the machine. Explain the difference between the electrical functions and the physical layout of electrical equipment.

3. On which side of the ladder diagram is the common conductor shown? How is it generally identified in both the drawing and the electrical panel?

4. The ladder diagram is always drawn as though the circuit is deenergized. How does this influence the way the diagram is read during testing when the control circuit is hot?

5. In what way can circuits in other parts of the electrical system influence the specific circuit on which you are working? What special awareness does this necessitate?

6. Why is the output device or component always the last component in the circuit with one terminal connected to the grounded common control wire?

7. Why is the grounded neutral leg of a control transformer never fused?

CHAPTER 4

On-Line Troubleshooting Overview

OBJECTIVES

After completing this chapter, you should be able to:
- **Describe and use basic on-line electrical troubleshooting.**
- **Understand the difference between on-line and off-line electrical troubleshooting.**
- **Practice safe on-line troubleshooting techniques in the field.**

EXAMPLE OF ON-LINE TROUBLESHOOTING

The simplest way to understand on-line troubleshooting is to see it done on a single ladder diagram line. Line 19 shown in Figure 4–1 has a number of switched contacts and should be relatively simple to understand. For the sake of this example, make a number of assumptions. First, assume that every mechanical and electrical contact is closed. That is, at the time of the test, CR10 will be deenergized, CR3 will be energized, S3 will be closed in the number 2 position, LS3 will be closed, CR7 will be energized, and CR6 will be deenergized. Also assume that wire 19, which is powering this circuit, is energized.

During this troubleshooting procedure, the machine is running and the controls are set so that solenoid F (mold close) should energize. If the machine were operating normally, the mold close function would sequence. However, for this example, there is an electrical problem in the machine, and the mold-close solenoid F is not functioning. Even though the hydraulic pumps are running, the machine is **stalled**.

When describing on-line troubleshooting, the term *stalled* is used to mean that the machine will not complete its cycle because of a control circuit malfunction. This occurs even though all controls are correctly set. Strictly speaking, the term *stalled* means that this control malfunction *causes the machine to stop and rest in an idle mode*. With hydraulic machinery, the pump motors continue running. Note carefully that we are defining a stalled machine as being one in an *idle* position. There are important safety considerations in identifying stalled equipment that will be discussed later in this chapter. In other troubleshooting instances, a control problem will cause a skip in a single machine function though the machine subsequently continues to cycle properly. When the machine does not stall, you need to modify the on-line troubleshooting technique. For this example, you will assume that the machine stalls and idles.

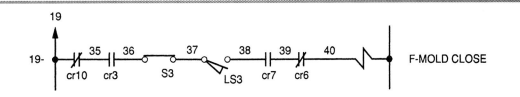

Figure 4-1 Diagram line 19.

In actual practice, you would now do a preliminary check of the machine. This will be covered later in Chapter 6. For this example, however, assume that you found no indication of a problem in your visual inspection of the machine. Your next step is to study the ladder diagram to determine the nonfunctioning area. Since the mold is not closing when it should, you can isolate solenoid F (mold-close) on line 19 as the problem area. You will need to make a careful distinction at this point between an electrical power component that is not functioning and a component that is faulty. The mold is not closing because solenoid F is not functioning. This is not saying, however, that the solenoid itself is at fault. The fault may be anything in the mechanical or electrical system that prevents the control voltage from reaching the solenoid.

The Troubleshooting Process

You are going to troubleshoot line 19 from Figure 4–1. However, your procedure will be different from that of most of your fellow electricians. They would probably stop the machine and check machine wiring for open circuits. For your on-line troubleshooting, you will do your testing from inside the electrical panel as shown in Figure 4–2:

Step 1. You will set the machine under power so that the mold should close. During your on-line troubleshooting, the motor will be running, the control circuit will be on, and the control selectors will be set so that solenoid F should be energized.

Step 2. You will start your actual diagnostic work by testing the circuit for continuity. Notice that you will be testing for continuity! Do not forget that the lines are energized. However, you will be using a **digital multimeter (DMM)**, which can test resistance and continuity with full control voltage on the line. As an alternative to the multimeter, you may be using a voltage and continuity tester, which is designed to read continuity on a live circuit. These versatile test instruments are the key to the speed of on-line troubleshooting. Specific meters will be identified for on-line troubleshooting in the next chapter. Refer to Figure 4–3 and Figure 4–4 for the type of test instruments capable of these measurements.

In Chapter 1, you studied some important safety procedures. Be certain that you understand all necessary precautions before working in a live panel. Do not attempt any testing until you know that you are adequately protected from all higher voltages, and that you will be using your test equipment on no more than 120–volt control circuits for any on-line troubleshooting tests.

Before going further, it is necessary to expand the earlier definition of continuity. A circuit with continuity provides a continuous path along which

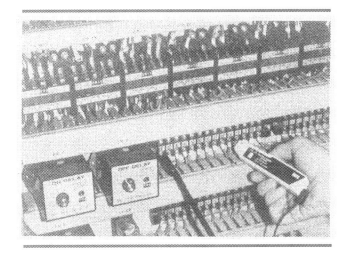

Figure 4-2 This illustration shows the pencil multimeter being used in an electrical panel. This particular meter can be set on a resistance function while testing an energized 120-volt control circuit.

46 CHAPTER 4 On-Line Troubleshooting Overview

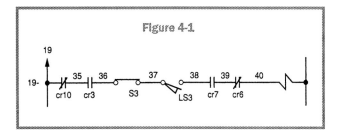

Figure 4-1

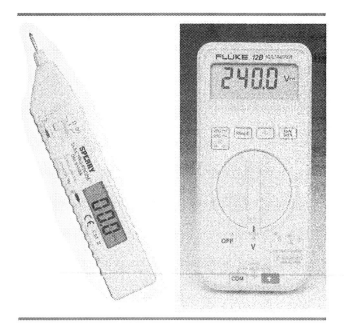

Figure 4-3 Two digital multimeters that can tolerate voltage while on a resistance setting. These meters are further identified in Chapter 5. *Courtesy of A. W. Sperry Instruments, Inc. and FLUKE Corporation.*

Figure 4-4 Greenlee Textron's model 6708 is an excellent combination test meter for on-line troubleshooting as described in this chapter. *Courtesy Greenlee Textron, Inc.*

an electric current flows. Stated differently, a conductor or a circuit has continuity when a current can flow through that conductor (or circuit) encountering no resistance significantly greater than the wiring and the devices of the circuit itself. A conductor has continuity. A nonconductor, for whatever reason, does not have continuity. Obviously, when you are testing a control circuit, it must have continuity before it can function. In order to do its job, it must carry current to the intended electrical output device or component.

You should also be aware that a standard **ohmmeter** cannot be used for continuity testing on a live circuit. *Unless your meter is specifically designed for live circuit testing on the ohm or continuity settings, do not attempt to read resistance or check for a completed circuit unless you know the circuit you are testing is deenergized.*

Step 3. Until you verify individual equipment on the machine itself, almost all of your testing will be done from inside the electrical panel. These techniques are explained in later parts of the book. Not only will you learn how to test remote machine components, but in Chapter 9 you will discover ways to actually verify mechanical functions—such as a solenoid valve's operation—from inside the panel.

Remember that the entire line 19 needs to act as a single conductor in order for solenoid F to function. This means that from any point on the left side terminal of the solenoid (wire 40) to any point along the entire circuit of line 19, you should be able to read continuity with your meter. However, since the solenoid is not energized, you will expect to find some point at which there is an open circuit. That, in fact, is the purpose of your test. You want to find the point along line 19 at which continuity is broken.

Actual Testing

You are now ready to start the actual testing from inside the panel. Most of the testing will be done from the numbered terminal blocks, though occasionally you will test from numbered wires on relays, timers, or other circuit components. For the purpose of this example, the ladder diagram has been altered. The complete diagram in Figure 4–5 shows relay CR5 as common to solenoid F when tr10 is closed between wire 40 and 41. If you were to test for continuity between wire 40 and the common wire, you would be testing a parallel circuit for both solenoid F and CR5. This parallel path would give you a reading even if solenoid F was removed from the circuit. In field testing, there are simple procedures that will allow you to verify each output

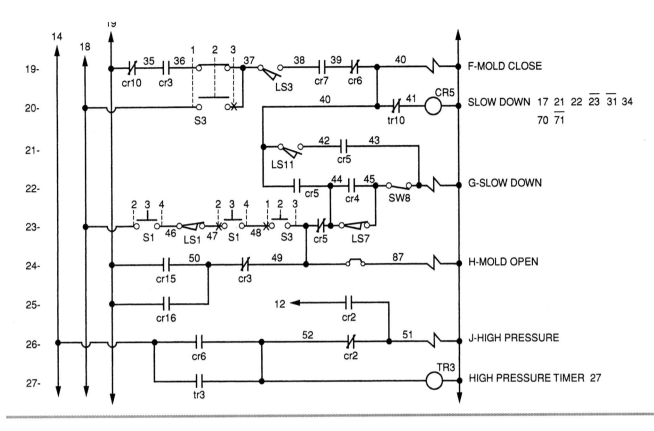

Figure 4-5 Diagram lines 19 through 27.

device or component separately. For now, however, use Figure 4–1 as though wire 40 is not common to any other parallel circuit. Your testing sequence is as follows:

Test Point 1
Your first and only voltage test is from wire 19 to the common grounded conductor on the right-hand side of solenoid F.

Test purpose. To verify the presence of a control voltage.

Test results. Your meter indicates 120 volts. Continue with the next test sequence.

Test Point 2
If you are using a protected digital multimeter as described earlier, change your meter setting to continuity. If you are using a tester that automatically ranges between continuity and voltage, you can continue without resetting. Test solenoid F for an open circuit by touching the leads to the common grounded conductor and wire 40.

Test purpose. To verify a complete circuit through solenoid F (mold close).

Test results. Beep = continuity. You now know that the solenoid and its circuits are operational. Your meter indicates continuity with a beep tone.

Test Point 3
Arbitrarily select a midpoint connection for line 19. Wire 38 is on the terminal block and is convenient to use, so test for continuity from the previous test point (wire 40) and the terminal block wire 38.

Test purpose. To test relay contact cr7, relay contact cr6, and all terminal connections between the two points.

Test results. Beep = continuity. You now know that all contacts and terminals are operational between wires 38 and 40.

Test Point 4
Now test the other half of the circuit for continuity by testing from terminal block wire 38 to wire 19.

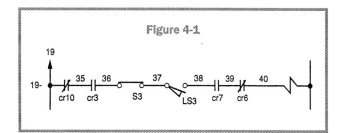

Figure 4-1.

Test purpose. To test LS3, selector switch S3, relay contacts cr3 and cr10, and all terminal connections between the two points.

Test results. 120 volt indication = open circuit. You now know that an electrical device or connection between wire 19 and wire 38 is faulty or open.

Note: The meter you are using may not represent a voltage value of 120 volts in its continuity setting. It may merely indicate the presence of a voltage.

Test Point 5
Test between wire 37 and wire 38 on the electrical panel terminal block.

Test purpose. To verify LS3 and its connections.

Test results. Beep = continuity. You now know that limit switch LS3 and its connections form a completed circuit.

Test Point 6
Test between wire 36 and wire 37 on the electrical panel terminal block.

Test purpose. To verify selector switch S3 and its connections.

Test results. 120–volt indication = open circuit. You now know that either the switch or the wiring is defective between these two points. You need to go to the control panel to verify it, since this part of the circuit is on the machine itself. However, before you do, make one last test from the electrical panel to make certain that there is only one area at fault.

Test Point 7
Test between wire 19 and wire 36 on the electrical panel terminal block.

Test purpose. To verify relay contacts cr10 and cr3.

Test results. Beep = continuity. Both relay contacts are operational. You now know that the fault is in selector switch S3 or its connections.

Test Point 8
Open the operator's control panel and locate the back of selector switch S3. Locate wire 36 and wire 37.

Test purpose. To verify that either the switch S3—or its connections—are faulty.

Test results. Touch the test leads to the switch terminals—not the wire ends. The absence of a beep verifies a faulty switch. Continuity indicates a poor wire connection.

You have now found the cause of the malfunction on line 19. It is either a faulty switch (S3) or a poor connection on one of the switch terminals. If you suspect a loose terminal screw, you would turn off the power and tighten the connections with a screwdriver. Remember, to this point you have been working on an energized circuit! Your tests will now verify that either the terminal contacts to S3 or the switch itself is faulty. The rest of the job will follow standard repair procedures. Turn the power off before doing any repair work.

At this point, you need to remember a very necessary safety precaution. During on-line troubleshooting, the machine is stalled in an operation mode while you are doing the testing. There are two safety concerns you must constantly be aware of:

1. You are working on live circuits. Be careful! In some cases, that may involve only a 24–volt control circuit. Personal risk is greatly reduced at the lower voltages. Nonetheless, carelessness still poses the potential of damage to the electrical equipment by short circuiting the control circuits with the test probes.

2. Be certain at all times that the machine is properly cleared so that it can continue its cycle with

neither personnel injury nor damage to the machine. This is necessary because in the process of testing, you may jiggle a loose contact and cause the machine to cycle. If you have taken the proper safety precautions, that in itself is a useful part of testing because jiggling a wire indicates that you have found the loose connection causing the problem. But if you have been careless, the machine cycle could cause personnel or equipment damage.

Precautions with Stalled Machinery

Remember that a stalled machine was defined as one with *a control malfunction that caused the machine to stop and rest in an idle mode*. The key to this definition is that the machine is *stopped* and *resting in an idle mode* because *the electrical controls* are not responding. Under no circumstances does this allow for testing on machinery that will not advance because of *mechanical* straining or binding.

With **hydraulic** or **pneumatic actuators** (cylinders) a great deal of energy may be stored even though a cylinder is not moving. A machine may stop functioning because the load limits have been exceeded, because something has broken or jammed, or because hydraulic or air pressure has been reduced (but is not absent). All of these conditions mandate a complete shutdown of the equipment and subsequent correction of the problem before any electrical troubleshooting involving hydraulic or pneumatic equipment is attempted.

Troubleshooting is never to be done on any system that has pressurized hydraulic or air lines leading to actuators (cylinders or any device capable of releasing energy or converting energy to motion). If, after carefully evaluating the system, it is determined that the equipment has not stalled because of jamming or any mechanical failure that could unexpectedly become free and move, and that the hydraulic or air pressure is normal for that operating condition, testing may continue if appropriate precautions are taken. These precautions must consider prudent and safe practices for the specific equipment and the manufacturer's recommendations for safe maintenance. Precautions would generally involve turning off hydraulic or air supplies and placing adequate blocking so that equipment will be prevented from moving (extending or retracting) or falling unexpectedly.

In a similar manner, all systems that might involve suspended weights or spring loads must be treated with appropriate caution. Be diligent in looking for potential hazards in charged systems. For example, a nitrogen accumulator on a hydraulic press could be lethal even though the hydraulic pump motor is off and all electrical controls are dead. Manually moving a hydraulic solenoid valve could activate a hydraulic cylinder through its complete stroke at maximum force.

If all other safety considerations have been met, testing *may* be done if hydraulic or air regulators are set so that pressure is greatly reduced in order to allow visual verification that actuators will "creep" with all loads removed. The same precaution would apply to weight or spring loads.

A Review of On-Line Troubleshooting

The following is a quick review of the on-line troubleshooting procedure you just finished.

First, you purposely put the machine in the running mode—most likely in its automatic cycle setting—and let it reach a point at which it would not continue its normal operating sequence. At the point at which the machine stalls, you can much more easily determine the area of fault. Checking a stalled machine will usually result in your being able to quickly isolate an output device or component that is not functioning, such as the mold closing solenoid in the example. However, even though it is not functioning, the device or component itself may be fully operational.

By leaving the machine stalled while it is still in the running mode, you can follow the ladder diagram and check the circuit for open conditions. This is conveniently done by subdividing the circuit and checking for continuity. At some point, you will find a section of the circuit that is open. At this point you can test specific devices on that circuit until you isolate an individual faulty device or connection.

For the sake of clarity, an important part of the testing procedure must be restated. In some cases, you will actually be doing your testing while the machine is in normal operation. In this case, it is obvious that all control settings, safety gates, and any other electrical equipment involved in the circuits you will be testing are in their normal positions. Thus, any testing will, in fact, be on a fully operational circuit. This will often be the case when you are checking a function that is not critical to the machine's main operation—such as an air blast ejection system. However, when you are operating a machine until it stalls, which means it has reached a point in its automatic cycle at which it will not continue because some mechanical or electrical function has not sequenced, *it must remain in exactly that operating mode during testing. That means nothing is changed on the machine or the operator's panel from the time the machine stalled until you are through with that portion of the on-line troubleshooting.* For instance, if you open a safety gate or turn a selector switch to the manual position, you will change the circuit in the panel, and you will no longer be doing on-line troubleshooting.

It is important that both you and the operators understand the procedure. If you are anticipating testing a piece of equipment, you need to carefully explain to the operators that when the machine ceases to function normally, they should call you without touching any control or part of the machine. Of course, you understand that this does not bypass necessary safety precautions. However, particularly when you are testing intermittent failures that you may not be able to reproduce at will, it is mandatory that you begin testing the electrical circuit in precisely the mode it failed.

Finally, you can do the majority of the testing from the electrical panel; you do not need to move from switch to switch on the machine. Testing individual switches or components is often done with considerable effort in exposing the switches and their wiring terminals. Only as a last step do you need to open operator control panels or limit switch mounting boards to verify your testing, or possibly to select between a small number of switches that do not have individual wiring to the electrical panel.

What did you gain by the procedure you just used? Speed! The entire procedure described in the troubleshooting example should have taken no more than 10 or 15 minutes for the actual testing, even if there was no prior knowledge of a malfunctioning selector switch.

COMPARING OTHER TROUBLESHOOTING METHODS

By this time you should have a basic understanding of how you will be testing with on-line troubleshooting. You will be testing for broken continuity (open circuits) in a circuit that is live. The machinery may even be in production. It is a live circuit because you want to test the circuit under its actual running conditions. That is the simplest and fastest way to find a circuit that is not operating normally.

Conventional Troubleshooting Methods

Conventional troubleshooting, however, is generally done off-line. In other words, the control circuit is deenergized. Every troubleshooter has a personal technique, but some form of off-line diagnosis is the most commonly used. The primary reason for deenergized control systems is the need to accommodate older-style volt-ohmmeters (VOMs). Until recently, it was not possible to use standard test meters for a continuity check on live circuits. A significant part of the check can, in fact, be done with a voltmeter. Nonetheless, it would mean switching back and forth between volt and ohm functions on the meter, while turning the electrical control circuit power on and off. Because of the potential of error and the resulting damage to equipment—and electricians!—many diagnostic procedures have been standardized as off-line.

Because the control circuit is off in conventional testing, "jumping" becomes a standard practice. If you want to make a continuity test with the control circuit deenergized in a circuit that has a relay electrically holding a contact closed, you must either mentally calculate the effect of that open contact, or you must physically complete the circuit with a jumper wire or manually hold a contact closed.

Advantages of On-Line Troubleshooting

In electrical troubleshooting, the effectiveness of a testing procedure has as much to do with the person doing the work as it does with the procedure itself. Other techniques besides those suggested here are effectively used every day by qualified electricians in industry. Nonetheless, with the solid-state tools that are available today, there are some distinct advantages in doing continuity testing while the equipment is on-line. The profitability advantage is almost always found in faster diagnosis and less downtime. The following are some of the advantages that come with the use of on-line troubleshooting, as compared with conventional troubleshooting methods.

1. *Downtime is reduced or eliminated.* Many times a troublesome electrical function will not disable equipment, but it nonetheless needs to be corrected. The easiest way to do on-line troubleshooting is from inside the electrical panel while the machinery is running. As always, poor safety procedures are not allowed. You are assuming that the electrical panel is properly constructed with no more than 120 volts as the control voltage, and that there are appropriate clearances from the higher voltages being controlled. The advantage is then very obvious, because this troubleshooting technique will often allow diagnosis without taking machinery off-line until the actual replacement of a faulty electrical component is made.

2. *Electrical components can be tested under load.* Shutting equipment down electrically for testing will always introduce some degree of variation in the circuit as compared with its operation under load. Loose terminals, pitted contacts, or weak contact springs may all operate differently with the power off. A loose terminal may easily carry the current from an ohmmeter, though it would not carry the actual control current. This is particularly true in the case of relay contacts.

   With the circuit live, the contacts will be properly sequenced and under normal load. It is only then that the relay contacts themselves can be properly included in the test. Most veteran industrial electricians have tested a deenergized circuit from the relay contact terminals to the rest of the circuit and found nothing wrong. Later, through much trial and error, they discovered that the cause of the problem was a single set of relay contacts that had been omitted because the continuity test was done from the relay terminal. Even if the contacts themselves are not faulty, it is much simpler to test a circuit with the contacts in their proper operating positions than to work through the wiring diagram while mentally figuring their sequence.

3. *Intermittent problems can be found more quickly.* Probably the hardest troubleshooting task for any electrician is an intermittent fault. It is usually the result of a loose connection or a failing relay or switching contact. Frequently, when doing off-line troubleshooting, the very procedure of turning off the power or removing the operating load from the circuit will offer enough variation to temporarily allow the device to carry sufficient current to indicate a positive reading with an ohmmeter. The result can be either a great deal of lost time trying to find the failing device or the expensive and unnecessary replacement of components in the circuit.

   With on-line troubleshooting, you set the conditions to do the testing on a live circuit that is in the actual process of failing. By allowing the machine to run until it stalls, you are testing all of the electrical components and circuits under actual operating conditions. If it is an intermittent failure and safety is not compromised, you can put the machine back into operation and wait until it fails. When the fault does occur, without changing any of the settings, you can proceed with the testing knowing that there is a circuit failure that will be evident. On the other hand, waiting by a running machine until it fails can be a time-consuming experience. It may be anywhere from a few minutes of operation to months. You are going to waste a lot of time trying to find an electrical fault that does not presently exist.

   Intermittent problems are often related to high resistance in relay or switch contacts. If you are using one of the recommended test meters capable of doing on-line troubleshooting, you may find a further diagnostic aid in its tone response. Some meters will vary their tone in the continuity setting when resistance is higher than 100 or 200 ohms. This description is generally true for the type of meter shown in Figure 4–4. The resistance value of closed relay contacts should be more constant. A faltering tone variation in the continuity setting may indicate potential high resistance, which is a clue to a failing contact. There is no rule to follow at this point, as each meter will respond differently. Get used to your own meter, however, and you may find that the tone will give you additional information.

4. *Electrical safety may be enhanced.* On the surface, it would appear that working on a live electrical circuit increases accident potential. That is certainly true when hazardous and sloppy procedures are used. However, there are areas in which more traditional troubleshooting can introduce its own hazard potentials. The most obvious is when jumper leads are used to complete tests. Even if the circuit is dead, it requires adding wires and alligator clips into a panel that was not designed for them. If the circuit is energized for temporary testing, there is a real hazard of shock or damage resulting from faulty connections. Jumping across contacts with a live circuit not only runs the risk of short-circuits, but also introduces the potential of closing circuits that jeopardize personnel or machine safety because they are out of sequence. Jumping, in effect, can remove electrical interlocks. If the machine is being tested using on-line troubleshooting techniques while it is in operation, that is not possible. With on-line troubleshooting, you are not altering or adding wiring to the manufacturer's circuit. Further, electrical safety functions will not be bypassed, nor can the machine be dangerously sequenced.

Thus, on-line troubleshooting has some decided advantages that evidence themselves in faster electrical diagnosis. At the same time, not only does it maintain electrical safety practices when properly done, but on-line troubleshooting may enhance safety by preserving the manufacturer's wiring and electrical sequences even during testing.

CHAPTER REVIEW

On-line troubleshooting is done with the machine or electrical equipment in the running mode. The equipment is either cycled in its normal operation mode, or it is operated until it stalls and will not continue to the next operational sequence. Troubleshooting is then done on the live circuit in an attempt to isolate a single malfunction in the control circuit.

On-line troubleshooting is ideally done as a continuity test on the live circuit. Test equipment that can tolerate control voltage during a continuity or resistance test is used. Troubleshooting is carried out by first testing large sections of the faulty circuit for continuity. When there is continuity, the procedure is carried out on adjacent sections of the circuit. When lack of continuity is indicated, circuit sections are subdivided until the subcircuit area is found where there is a break in continuity.

Safety is a major concern of any troubleshooting procedure. In the case of on-line troubleshooting, in addition to the care taken while working on live circuits, there is the need for special precaution to avoid personnel or machine damage caused by inadvertent machine cycling during testing. On-line electrical testing must never be done on a machine that is mechanically bound or jammed. Any mechanical loads that might move or be released must be safely secured before proceeding with any testing.

On-line troubleshooting offers some definite advantages over conventional troubleshooting. Downtime can be greatly reduced, or, in some cases, eliminated entirely. Electrical devices can be more accurately tested under actual load. Intermittent problems can be found more easily under live circuit operating conditions. Finally, electrical troubleshooting safety may actually be enhanced because the need for electrically modifying or jumping circuits is eliminated.

THINKING THROUGH THE TEXT

1. How are machine controls set for on-line troubleshooting?

2. Define *continuity*.

3. Two safety precautions are mentioned that must be uppermost in your mind when you are conducting on-line troubleshooting tests. What are

these two precautions? What are the possible consequences of their violation?

4. On-line troubleshooting is often done when the equipment is "stalled." Describe what is meant by a stalled machine according to the definition of the text. In what position are the controls when the equipment is stalled? What is the difference between a *stalled* machine and a *jammed* machine?

5. What is "jumping"? How is it used? How might jumping increase the potential of danger while troubleshooting?

CHAPTER

5

On-Line Troubleshooting Tools

OBJECTIVES

After reading this chapter, you should be able to:
- **Understand the place of testing procedures as well as the test equipment itself in producing effective on-line troubleshooting.**
- **Select instruments that will optimize your electrical troubleshooting efforts.**
- **Understand the differences and applications of analog volt-ohmmeters (VOMs) and digital multimeters (DMMs).**
- **Define a true RMS voltage.**
- **Use non-contact meters in on-line electrical troubleshooting.**

THE IMPORTANCE OF TEST EQUIPMENT

Helping you improve your troubleshooting speed and accuracy is a major objective of this book. Your troubleshooting effectiveness will certainly be a function of your testing procedures. However, it will also be a result of the electrical testing equipment you are using. The explosion of new electronic diagnostic equipment is bringing the potential of amazing speed and accuracy to the maintenance electrician's normal working day. As this chapter—and the remainder of the book—is developed, you will be introduced to some of these test instruments. Figure 5–1 shows some of the electronic test equipment that is now available.

Electronic Test Equipment Makes New Methods Possible

In spite of the new equipment that is now available, many of the testing procedures for the plant maintenance electrician still unnecessarily rely on the methods previously required by **moving-coil** volt-ohmmeters. Continuity readings are still done as though the circuit must be de-energized.

More importantly, the full capability of much of the new equipment is not being satisfactorily used. Test instruments are available today that can perform a full complement of waveform and diagnostic tests. Too often, however, they are used in plant maintenance work as little more than electronic volt-ohmmeters.

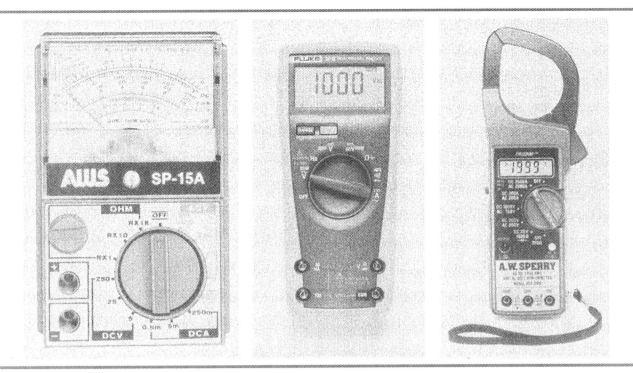

Figure 5-1 This figure illustrates common electrical test equipment that a plant electrician would use. A basic analog volt ohmmeter (VOM) is shown on the left, a digital voltmeter (DMM) is shown in the center, and a clamp-on ammeter is shown on the right. *Courtesy of A. W. Sperry Instruments, Inc. and FLUKE Corporation.*

The value of this new test equipment goes beyond mere speed. Maintenance work can now be done with less machinery disassembly, less contact with live circuits, and less time spent working in the electrical panel, all of which lead to greater personnel and equipment safety. The use of this new generation of test equipment can often reduce the hazards associated with electrical maintenance work.

Choosing Test Instruments

You will see numerous references to specific brand names and model numbers of electrical test instruments in this chapter and the remainder of the book. In some instances, the tests being described can best be done with specific instruments. Generally, however, describing named meters will allow you to locate other manufacturers' equipment that will accomplish similar results.

Nonetheless, each of the named test instruments is recommended as an excellent meter for the job for which it is described.

CONTACT AND NON-CONTACT MEASUREMENTS

Electrical measurement is essentially achieved in one of two ways. The most common measurement method is contact testing. In contact testing, leads or probes of the meter make electrical contact with the conductors of the circuit under test. The circuit may be either energized or deenergized. In contact testing, the circuit of the meter is electrically common to the conductors of the circuit. The second method is non-contact testing. In non-contact testing, the internal circuit of the meter is isolated from the conductors of the circuit under test. In this case, the meter is sensitive to an **electromagnetic field** surrounding the conductor. The current may be induced in the meter by transformer action as is done with a **clamp-on ammeter**, or the meter may be sensitive to capacitance changes or other properties of an electromagnetic field.

Contact Testing

Volt-ohmmeters (VOMs), which are moving-coil **analog meters** with needle and scale indicators, and digital multimeters (DMMs) or digital voltmeters

(DVMs), which are electronic meters with solid-state circuits and a digital readout display, are probably the most recognizable instruments used for contact testing. Both of these meter types are shown in Figure 5–1. All standard tests using these instruments are done by touching their leads to the actual electrical circuit. The electrical circuit, whether it is an energized lead or a chassis ground, is thus common to the internal circuit of the VOM or DMM. This is true of any voltage, resistance, or **milliamperage** measurement done with a VOM or DMM. However, some multimeters have temperature or clamp-on ammeter attachments that fall outside of the normal use of a VOM or DMM as a contact testing instrument. Many other tests are also made through direct contact with the circuit being tested. Capacitor testers, continuity testers, **phase rotation** indicators, and hand-held **oscilloscopes** or **harmonic** meters are all examples of test equipment that require direct contact with the circuit during the test.

Non-Contact Testing

The clamp-on ammeter, on the other hand, is probably the best known example of a non-contact test instrument. A clamp-on ammeter is shown in Figure 5–1. In order to take a reading, the ammeter's laminated steel jaw is clamped around an alternating current conductor. No physical connection with the live circuit is required. A current is induced in an internal winding in the ammeter itself, though the ammeter's circuit is completely isolated from the circuit being tested. The ammeter is really a transformer secondary connected to the instrument's moving coil or digital readout, which is calibrated to show ampere values. The finely wound coil at the bottom of the jaw is the transformer secondary, and the conductor clamped in the jaw is the **transformer primary**.

Many other types of testers have been built that use some form of induced or capacitive interaction with an insulated conductor so that a value in the conductor can be read without electrical contact with the circuit in the conductor. Enhanced safety and the ease of use are the primary considerations behind the development of non-contact test equipment.

Although this section has introduced the two types of instruments as though all meters are exclusively contact or non-contact, there are, in fact, a number of combination meters on the market. Probably the most frequently seen is the non-contact clamp-on ammeter with a built-in DMM requiring contact meter leads as seen in Figure 5–1. There are other interesting combination meters such as TIF Instrument's Tic Tracer 300 shown in Figure 5–2. The primary instrument is a non-contact proximity meter that detects the presence of an AC voltage. Built into the meter, however, is circuitry requiring leads for contact continuity testing.

Always Verify Your Meter

Before going further, a procedure that applies to almost all testing described in the remainder of this book must be established. After turning a meter on, or after changing a meter's range, always verify that the meter is properly working on a known source. For instance, when you are testing for voltage, place the probes on a known energized wire to make certain that the meter responds properly. For obvious reasons, a malfunctioning meter can create an extreme hazard if it fails to indicate a voltage and you presume the conductor is deenergized.

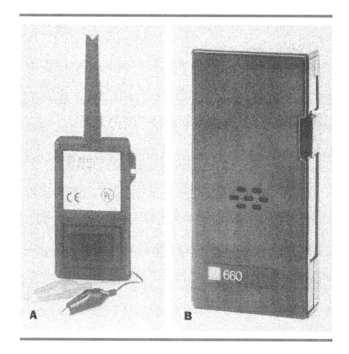

Figure 5-2 Two combination test meters from Tif Instruments, Inc. (A) The tic Tracer 300 is a proximity test meter with a continuity function. (B) The TIF660 is a capacitor tester with a continuity function. *Courtesy of Tif Instruments, Inc.*

You must appropriately match the test voltage to the setting. If your meter is set on a 600–volt range, it is entirely permissible to touch the probes to a 120–volt control voltage to verify the meter before using it to test a 480–volt contact. However, you would not reverse that procedure and use a 480–volt source to test a meter set on a 200–volt range. Nor would you test the meter on a 480–volt source while in the 600–volt range and subsequently switch the meter to the 200–volt range to verify that a control circuit was deenergized. Other ranges use modified verification procedures. Generally, any ohm range can be verified by merely touching the two probes together. *Remember, a conventional ohmmeter must never be placed on an energized circuit.*

It is particularly important that non-contact proximity testers be verified before working on supposedly deenergized conductors. Each manufacturer's proximity tester works differently. However, most proximity testers will respond when tapped briskly against the hand. If the proximity tester uses a pocket clip as the "on" switch, do not release the pocket clip between the hand test and the wire test.

The primary concern in verifying any meter is safety. *Do not depend on a meter reading to indicate a deenergized circuit without verifying the meter in the same setting on an energized circuit.*

MOVING-COIL METERS

Before examining digital and solid-state equipment, stop for a brief look at the earlier workhorse of electrical testing equipment. The first multimeters used a needle that swung an arc across a scale. Figure 5–3 shows the internal construction of an analog moving-coil meter, technically known as a permanent-magnet moving-coil meter.

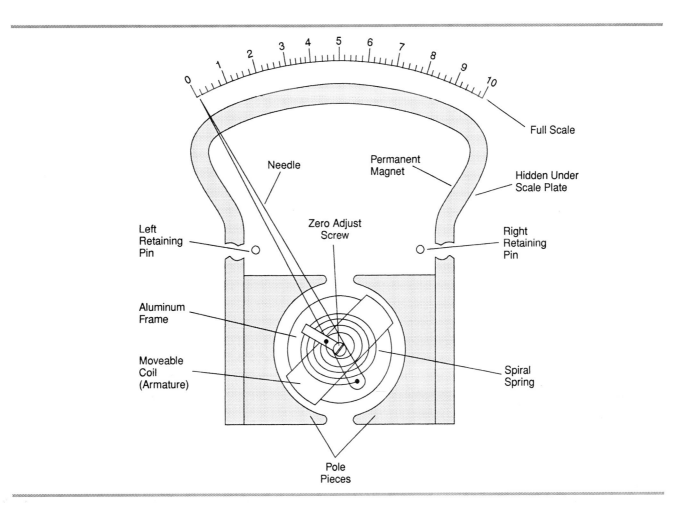

Figure 5-3   The internal construction of a D'Arsonval moving-coil meter. *Courtesy of Master Publishing, Inc.*

Even though the moving-coil meter has been largely replaced by the DMM in industrial electrical troubleshooting work, a review of its basic function is in order. First, it is worthwhile to understand how the meter works because moving-coil meters are still used in many lower-cost test instruments. Secondly, moving-coil meters are also used as panel meters, readouts on welding equipment, and as machine rate indicators, and the like. Understanding their function also aids in proper care to ensure their accuracy. Finally, in some specialized applications, the pointer movement itself indicates a condition that is not as easily detected on a DMM readout. For example, a very sharp voltage rise and fall may not show on a digital display, whereas the mechanical dampening effect of a VOM will cause the pointer to noticeably swing.

Moving-Coil Meter Construction

A moving-coil meter can be compared to a permanent-magnet direct current motor. Like a direct current motor, the meter movement has a fine wire-wound **armature** that is free to rotate. Unlike a motor, however, the armature has a fixed rotation limit of about 90°. In the most common meter arrangement, the electrical current is commutated to the armature through two spiral springs rather than through brushes as in a direct current motor. When a direct current of the correct polarity is applied to the armature, the armature will rotate clockwise. However, since the armature is opposed by the spiral wound springs, the amount of rotation is dependent on the magnetic field strength of the armature. The field strength is a function of the applied current. As the current is increased, the deflection of the needle is increased relative to the stiffness of the coil spring.

A moving-coil meter is sensitive to current, not voltage. However, the value imposed on the moving coil in almost all VOMs is a voltage value. A short history note will help. One of the important discoveries in the development of electricity was that of electromagnetism by Hans Christian Oersted in 1819. In 1820, he published a short paper describing the magnetic field surrounding a conductor. It was the application of this discovery that allowed the development of induction and magnetic equipment such as motors and transformers. Oersted accurately described the force between the conductor and the magnet as being a function of current and not voltage. How, then, does the moving-coil meter vary with a change in voltage? The answer is an application of Ohm's law, which states that the current in an electrical circuit is directly proportional to the voltage in the circuit and inversely proportional to the resistance in the circuit. Therefore, if the voltage value increases, the current value will also increase. Thus, the moving-coil meter can respond to voltage because an increase in voltage will increase the current, which, in turn, increases the magnetic flux, causing the needle movement. For the sake of accuracy, this chapter generally refers to the moving coil as responding to current even though the value being measured is, in most cases, voltage. This explains, however, why the circuit for the lowest milliamperage range on the volt-ohmmeter schematic in Figure 5–4 flows directly through the meter.

A high-quality VOM uses a very sensitive moving coil. Calibrated springs and a balanced coil and needle result in a linear deflection of the needle across the scale as the current in the coil is increased linearly, meaning that the deflection of the needle is in exact ratio to the applied current. The coils are wound around an aluminum frame, which provides a dampening action on the needle movement. The frame, in effect, is a single-turn conductor. As the shorted conductor moves in the magnetic field, a voltage is induced in the aluminum frame. The induced voltage requires work energy, which, in turn, dampens the movement of the needle, keeping it from overshooting and oscillating.

A very small direct current in the range of 50 to 100 milliamperes (mA) is required for full-scale deflection of the moving-coil meter. To appreciate the sensitivity of the coil, look at the lowest direct current voltage range on your meter. Since even the lowest direct current range in most meters has **series resistance** in the meter's circuit, you know that the actual maximum allowable voltage to the coil is less than that of the scale value. In most cases, full-scale deflection in the lowest range would require no more than 2 volts. The sensitivity of a meter is often a measure of its quality. The more sensitive the meter, the less current that is required for full-scale deflection. Greater sensitivity is achieved with more turns

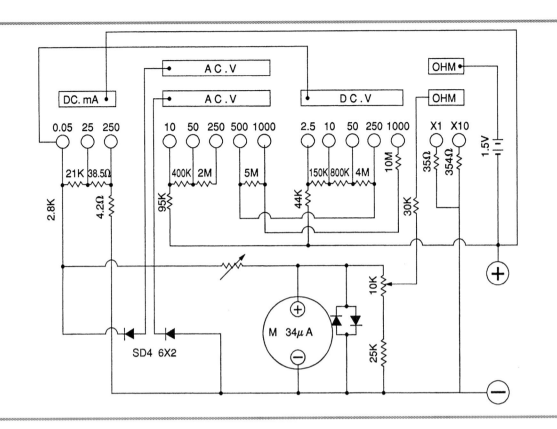

Figure 5-4 The circuit schematic of a moving-coil volt-ohmmeter (VOM). *Courtesy of A. W. Sperry Instruments, Inc.*

of finer wire on the armature winding. Greater meter sensitivity means that the meter is putting less load on the circuit it is testing, which results in a more accurate reading.

All meter operations, whether in the measurement of alternating current or direct current volts, resistance, or current, are a function of an applied voltage to the armature of the moving coil. The maximum allowable voltage is approximately 0.05 volt (50 **millivolts**) direct current. The meter selector switch adds resistance to the meter circuit so that it can reduce the voltage to safe limits for the moving coil. As noted earlier, this reduction of voltage through series resistance has the effect of reducing the coil current. In addition, in the alternating current ranges, a bridge rectifier is also put into the circuit to change the alternating current to direct current. In the case of the resistance ranges, a battery circuit is included, which provides the power for the moving coil. Ammeter ranges comprise shunt resistors that allow a voltage drop across the resistance elements to be measured. Figure 5-4 shows a typical VOM circuit schematic.

Accurately Reading the Moving-Coil Meter

Meter accuracy is dependent on two factors, one of which you can control on the meter itself. The first factor, which affects the VOM's accuracy, is the quality of its design and construction. You will get what you pay for, but beyond that, you cannot increase an existing meter's internal accuracy. However, in using a meter, you can control the precision of its measurements by properly selecting the range on which you take a reading. For a given VOM, the accuracy will be stated as a percentage of the full-scale reading. For example, if you have a meter rated at 1 percent accuracy, you would get the following results for a 2–volt value on three different scales. If you were reading 2 volts on a 10–volt scale, the accuracy would be +0.1 volt (5 percent). If you took the same 2–volt reading on the 5–volt scale, the accuracy would be +0.05 volt (2.5 percent). Finally, if you took the 2–volt reading on the 2.5–volt scale, the accuracy would be +0.025 volt (1.25 percent). Thus, the most accurate readings are those that are closest to the full-scale value of the meter. This is true of any value, whether it is volts, ohms, or amps.

Using a Moving-Coil Meter

You should be aware of several precautions in handling and using a VOM. When you are measuring unknown values of voltage or current, always start at the highest range. Do the first measurement of an unknown AC voltage with the function selector switch set on 600 volts AC or higher. Subsequently, you can set the function selector on the appropriate range after first reading the voltage on the 600–volt scale. This protects the meter from excessive voltages. Do not change the function selector switch while the meter is under load. Induction in the circuit you are measuring may create a high-voltage transient when the meter function selector switch is moved.

When measuring direct current voltages, start at a high value and momentarily touch the probes to the circuit. Watch the direction of the needle's deflection. If it is in the reverse direction, change the polarity of the probes before continuing the testing. Setting the meter on the highest range protects the coil from damage if it receives full voltage in a reverse-polarity mode. Probably the most important operating rule for all VOMs (and most DMMs) is to use the resistance settings on deenergized circuits only. There are frequent references to continuity testing on live circuits in this book. This must never be done with a moving-coil VOM in a resistance range. Reading a voltage with the meter on a resistance setting will either blow the **overload protection**—usually a replaceable fuse—or destroy the meter if it is not protected.

When you have finished using a VOM, turn the function selector switch to the "off" position or the highest voltage setting. Never leave a VOM in the resistance settings, both because there is danger to the meter if the probes accidentally come in contact with a live circuit, and because it will drain the batteries while in storage.

In spite of the emphasis in this book on newer electronic test equipment, the VOM is still a valuable instrument. When properly used, it is capable of a wide range of troubleshooting procedures and is a valuable tool for the maintenance electrician. In some instances, the actual pointer movement will indicate to the experienced electrician information concerning the circuit under test.

Troubleshooting with a Multimeter

From this point on, it is assumed that you will be troubleshooting with a digital multimeter (DMM) or digital voltmeter (DVM). Only occasional reference will be made to a volt-ohmmeter (VOM) when applicable. However, most comments regarding the DMM would also apply to a VOM.

Testing for voltage or low-resistance values is the most common industrial troubleshooting work done with a DMM. The VOM X1 resistance scale, or the DMM 200–ohm scale, is most frequently used to determine whether a circuit is open or has continuity. In these cases, the reading will be either **infinity** for an open circuit, or a reading of a very few ohms for a completed circuit.

Internal DMM ammeter milliamp ranges are primarily used in electronic rather than electrical work. Industrial electrical tests usually require ammeter ranges into the tens and hundreds of amps. For example, a 10-horsepower, three-phase motor running at full load on 240 volts would draw approximately 18 amperes. The starting current would be approximately six times the full-load running current or 108 amperes. Needless to say, the internal ampere ranges of multimeters cannot be used on these high values. In the industrial setting, ranges of 0.2 to one **ampere** are frequently used for measuring coil current. However, the clamp-on style of ammeter serves this function much more readily and is safer than disconnecting leads so that the DMM can be placed in series with the coil.

Most DMMs perform the same testing functions as their VOM counterparts with a lower percentage of internal error and greater reading precision. The reading precision is greater because the eye is incapable of distinguishing small interval variations of a needle on a scale. In contrast, the DMM will display the number in tenths—and even hundredths—of the scale value. Unless specified otherwise, the same meter precautions for measuring deenergized circuits while in the resistance ranges apply. That is, though on-line troubleshooting testing for continuity is described in this book, it is possible only on meters appropriately designated for use on live circuits while in the continuity range. *Do not assume that any DMM will safely perform resistance or continuity measurements on a live circuit unless specifically stated.* "Overload protection" means the meter has a fuse or internal circuit-breaker. It does not mean the meter is designed to operate with a voltage on the conductors during resistance testing.

The primary tests in which a DMM would be used in electrical troubleshooting work are as follows:

1. *Voltage value tests.* With the selector switch set on the appropriate (or higher) voltage range, the test leads are physically held against any two conductors while an alternating current or direct current voltage value is read. This test is concerned with an actual value; therefore the DMM digits are read with precision. An example would be a comparison between the phase leads on a three-phase motor to determine the voltage imbalance between phases. That test would be conducted by touching the leads to each of the conductor pairs and reading the numeric values. For example, phase A-B might read 236 volts, phase A-C might read 237 volts, and phase B-C might read 232 volts. You would know that the maximum imbalance was 5 volts between phases A-C and B-C.

2. *Voltage status tests.* With the selector switch set on the appropriate (or higher) voltage range, the test leads are physically held against any two conductors to determine the presence or absence of a voltage. A specific voltage value is not the concern. Rather, the mere presence (or absence) of a voltage indicates a status. An example would be a test across the two secondary terminals of the electrical panel's control transformer to verify a voltage to the control circuit. If VOM leads are touched across the two terminals, a needle deflection to the right would indicate that control voltage is present. The absence of needle movement would indicate an open circuit in the control transformer or in the conductors feeding the control transformer. A DMM is used in the same way, though the digital readout is watched to determine if the digits begin counting to a significant voltage value.

3. *Live circuit continuity tests using voltage.* With the selector switch set on the appropriate (or higher) voltage range, the test leads are physically held against two points on a common (energized) conductor. If the circuit is unbroken between the two test points, the voltage reading will be either zero or a very low value if there is a slight voltage loss in the line. A zero voltage reading is an indication—though not an absolute verification—of continuity.

On the other hand, if the circuit is open between the two test points and the conductors are energized at both ends of the circuit, the DMM readout will indicate a full line voltage. A ladder diagram line can be checked for continuity by touching the two leads to various points along the common conductor. A voltage reading would indicate an open circuit. For example, if the circuit of line 19 in Figure 5–5 was being tested for continuity, the test leads could be placed on wires 38 and 40. A full voltage reading would indicate that some part of that circuit was open. The open circuit would be either cr7 or cr6.

However, there is a potential of false readings. A zero voltage reading may indicate continuity. On the other hand, a zero voltage will also be obtained if there are two open circuits in the same conductor. If there are any other open circuits outside of—but common to—the two points being tested, the voltage difference between the two points will still be the same. For example, if the DMM leads are touched to wires 38 and 40, but LS3 is open, the voltage reading will be zero. Thus, for an accurate test, it would be necessary

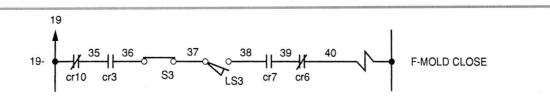

Figure 5-5 Diagram line 19.

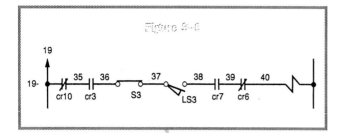

Figure 5-5

to verify that there was a full potential voltage between *both* test points and the common grounded conductor. That is, if wires 38 and 40 indicate a zero voltage when the leads are touched to them, the leads must then be touched between wire 38 and ground and wire 40 and ground to verify that, in both cases, there is a full voltage. If in any of the two successive voltage tests there is no full line voltage, the original continuity test is invalid. If LS3 were open, neither wire 38 nor 40 would show a voltage to ground, though the test between 38 and 40 would be zero. The advantage of on-line continuity testing, as already described, over voltage and continuity testing should be obvious. The continuity test is a measurement of the true state of the conductor, whereas the voltage test is affected by other variables. There can be considerable time loss in verifying these additional variables.

4. *Dead circuit resistance tests.* After verifying that all parts of the circuit to be tested are deenergized, the meter is set on the lowest resistance range for continuity testing. For the resistance tests, the meter's leads are physically touched to the conductor at the extreme ends of the section of conductor to be tested. A high-value reading indicates infinity, or an open circuit. A low-value reading indicates continuity. As an example, the circuit of line 19 in Figure 5-5 could be tested using a DMM as a continuity tester. The resistance setting would be used to read continuity by touching the leads to any two test points on the conductor.

Because of the lack of power in the control circuit, all of the relay contacts have relaxed to their deenergized position, which makes continuity testing more difficult. For that reason, the following example will be given demonstrating a complete testing procedure for line 19 in Figure 5-5. After you have verified that the power is off, place the leads of the DMM on wire 19 and wire 35. This will verify the conductivity of the normally closed contact cr10. The next contact, cr3, cannot be tested, since it is open. Therefore, the leads will be placed on wires 36 and 38. This will verify the conductivity of switch S3 and limit switch LS3, which should be closed because the machine is set to cycle. The final continuity test is between wire 39 and the common grounded conductor. This will verify the conductivity of cr6 and the mold-close solenoid F. Relay contact cr6 can be tested on a deenergized circuit since it is a normally closed contact. On the other hand, cr7 is bypassed because it is open when the circuit is deenergized.

If a break in continuity was discovered in the test completed so far, the cause can be isolated with further continuity tests. However, if the problem was not found, then there are still two additional relay contact areas that must be tested. They are the two normally open contacts. They can be tested in one of three ways. With the control circuit still in the off position, place the two lead probes on the relay contact terminals (wires 35 and 36 for cr3 or wires 38 and 39 for cr7). Most control relays have some provision for manually closing the relay. If you push a screwdriver against the mechanical relay push button, the DMM should read continuity if the relay contact is operational. The second way the contacts can be tested is to again set the DMM for a voltage reading, energize the control circuit, and set the machine in a stalled position so that line 19 should be functioning. With this setup, you can check for a voltage across either set of contacts as described under live circuit continuity tests.

The entire circuit can also be tested for continuity by jumping past the two normally open relay contacts. Jumper wires would be placed from wires 35 to 36 and from wires 38 to 39. Electrically, the two relay circuits are completed so that a continuity test can be taken on the entire circuit from wire 19 to wire 40. Aside from the safety consideration of using jumpers mentioned in Chapter 4, there is also the practical consideration that you may have bypassed the actual problem. That *will*, of course, be a clue as to what you need to check further, but it will add time to your troubleshooting procedure.

You should be able to see the advantage of increased speed and accuracy inherent in on-line troubleshooting as compared with the troubleshooting procedures using a multimeter, which requires that the circuit be deenergized for resistance tests. On-line troubleshooting can be summarized as having two advantages:

1. You can test the entire circuit in its operational mode. Because the relays are electrically cycled, you do not need to use jumpers or bypass test points.
2. You are not introducing variables that are outside of the normal circuit operation. A relay contact may read continuity when you have closed it manually with a screwdriver, but you have not proven that it does not have high (or infinite) resistance when it is operating under load.

DMM Terminology

Some basic terms need to be defined before going further. The first two terms are *function* and *range*. When used of a test meter, the term *function* identifies the type of test being performed. A basic DMM will usually have at least five functions, which include direct current voltage, alternating current voltage, resistance in ohms, direct current milliamperes, and alternating current milliamperes. Within each function, there may be various *range* settings. A typical mid-priced DMM may include direct current voltage ranges of 2, 20, 200, and 1,000. The alternating current voltage ranges would be 2, 200, and 750. The resistance ranges in ohms would include 200, 2000, 20K, 200K, and 2000K. The same meter would also have alternating current and direct current amperage ranges. How DMM scales are read are explained under the next heading.

Another term you will encounter is *autoranging*. Many DMMs are constructed so that a single function setting is used for all ranges. In this case, the meter is set on a single function and will automatically display the voltage or resistance reading of any value by shifting the decimal place. An autoranging meter is both simpler and faster to use because you do not need to start unknown tests on high range settings. It also protects the meter for the same reason since the meter cannot be damaged because it was used on a high voltage in a low-range setting.

A fourth term is *auto selection*. An auto selection meter automatically selects the function. There are two types of auto selection meters. We will identify the first type of meter as *high resolution*, meaning that true voltage or ohm values are numerically displayed.

Very few high-resolution, auto selection meters are on the market. Interestingly, however, FLUKE supplies two low-priced, general-purpose meter series with this capability. FLUKE identifies this function as VoltCheck™. The first is the FLUKE 7 series including the 7-300, which tests up to 300 volts, and the 7-600, which tests up to 600 volts. The second is the FLUKE 10 series including both the 12B and the 16, which use the VoltCheck function in the ohm range. In all of these meters, the probes can be placed on any alternating current or direct current voltage up to the rated value of the meter or on any resistance not exceeding 400 ohms, and the meter will automatically select both the function and the range before displaying the value. Both the FLUKE 7 series and 10 series meters are shown in Figure 5–6.

For the purpose of on-line electrical troubleshooting, this is an ideal instrument because it automatically selects the value needed at each test point. Because it displays a true ohm value, it can verify the difference between a high resistance contact of 20 or more ohms and a properly seating contact of one ohm or less. At the same time, it will give a true voltage value if the circuit is open.

There is a more common auto selection meter identified as a *low resolution* lamp meter. Typically, this meter uses **light-emitting diodes (LEDs)** and audio signals to indicate predetermined voltages and continuity. There are a number of these meters on the market, and they are excellent for on-line troubleshooting. Their convenient shape makes them easy to use because the meter and the probe are combined, facilitating two-hand use in an electrical panel. My preference in this category of instruments is the Greenlee 6710 because it has a digital voltage display rather than LEDs. The FLUKE T2 is a similar meter with the advantage of a CAT III protection rating. The Greenlee and FLUKE meters are shown in Figure 5–7.

64 CHAPTER 5 On-Line Troubleshooting Tools

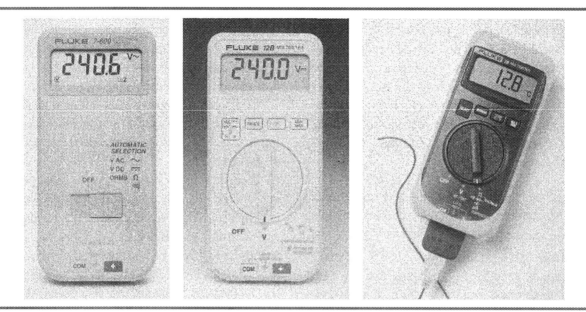

Figure 5-6   FLUKE Produces a number of inexpensive meters that display either a voltage or a resistance value without changing function. Their series 7 automatic range selection meters (left) automatically select and display either a voltage value or a resistance value. The 12B (center) and 16 (right) have more functions and are better choice for on-line troubleshooting. They actually display a resistance value in the presence of not greater than 4.5 volts, which allows reading relay contact resistance under operating conditions. *Courtesy of FLUKE Corporation.*

A final type of DMM we are defining is a *protected ohmmeter*. This meter has protection on the resistance function that allows it to be momentarily used on a 120–volt control circuit. In contrast to the high-resolution function meter above, these meters will not give a true voltage reading. However, when they are set on the resistance function, they will give a true ohm value across a test point. In the same setting, when an open circuit carrying a potential difference not greater than 120 volts is encountered, they

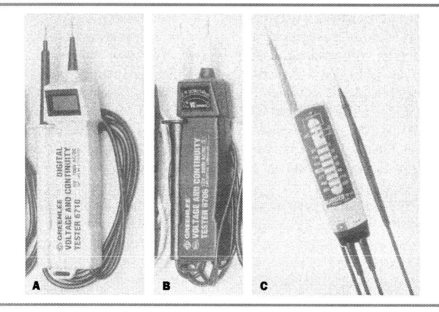

Figure 5-7   Three excellent combination low-resolution lamp meters for use in on-line troubleshooting. (A) The Greenlee 6710 with numeric voltage display. (B) The Greenlee 6706 with LED display. (C) The FLUKE T2 with LED display. *Courtesy of Greenlee Textron and FLUKE Corporation.*

will give a visual and/or audible signal that a voltage is present. In this way, they can be used to determine either a low value resistance (continuity) with a displayed value, or the presence of a voltage without changing functions. An example of a high-quality "pencil" multimeter with this function manufactured by A. W. Sperry (AWS) is shown in Figure 5–8.

Many DMMs may actually be capable of functioning in this way. Carefully reading the operator's manual may indicate that they can sustain short duration contact with up to 250 volts while in the resistance ranges. However, do not presume that degree of protection for any meter when the manufacturer does not make the claim in writing. Consequently, only those meters that state that they are capable of this overload usage are listed. Remember that improper use of a meter will certainly void its warranty, to say nothing of introducing an electrical hazard. My preference in this category is the FLUKE T5–1000 shown in Figure 5–9, followed by any of the pencil multimeters.

The AWS DM-6593A multimeter shown in Figure 5–8 is an excellent example of a pencil meter that can be used for continuity testing on an energized 120-volt control circuit. In the section covering resistance measurements, the operating instructions manual states, "Protection is provided up to 250 Vac/dc, but readings will not be accurate." When this meter is used in the ohm function for numeric display of ohm values, it will indicate an overload (without an audible tone) on the display when a voltage is encountered. If the meter is locked into the continuity setting, it will audibly indicate continuity, and will squawk when a voltage is encountered. The meter is comfortable to hold and easy to switch between functions for confirmation of continuity, ohm values, or voltage. Its CAT II classification is particularly important. It has a safety overload feature, which is its only disadvantage. It requires about 20 seconds to automatically reset after displaying a voltage overload in the ohm setting. However, this can be immediately cleared by switching the meter off and back on.

Note: **Impedance (Z)** *is the total resistance of a circuit to the flow of an AC current. Typically, the resistance ranges of a meter have considerably lower impedance than the voltage ranges. Any meter placed across a live voltage in its low impedance ohm function must dissipate the internal*

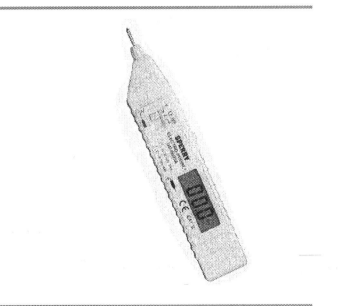

Figure 5-8 An example of pencil multimeter (A. W. Sperry's Electro-Probe DM-6593A digital multimeter). This multimeter package is extremely compact, yet it has a full range of test functions with a CAT II rating. This multimeters can be used on 120 volts in the resistance range. *Courtesy of A. W. Sperry Instruments, Inc.*

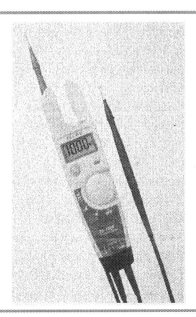

Figure 5-9 Two models of FLUKE's T5 combination meter are available with upper voltage limits of 600 for 1,000 volts. This small, easy-to-use meter measures both AC and DC voltages, resistance to 1,000 ohms with the ohm range protected to 600 or 1,000 volts, and a 100-ampere ammeter. *Courtesy of FLUKE Corporation.*

heat caused by the current flow. The FLUKE meters mentioned in this section can tolerate continuous use in this mode. Other meters may not tolerate the heat rise of prolonged voltage while in the resistance range. Repeatedly defeating an automatic reset may result in excessive heat in the instrument's circuit board, which can destroy the meter.

Reading VOM and DMM Scales

VOM and DMM scales are usually read directly for voltage values. This means that a VOM will read 120 on a scale, and a DMM will read 120.0 on a numeric display for 120 volts. The difference, however, is that the DMM moves the decimal point for various range selections, whereas a VOM may use different scales on the meter face to compensate for the meter's mechanical characteristics. A typical DMM may have four direct current voltage ranges of 2000m, 20, 200, and 1,000. In the first range, this meter will read up to 1.999 volts. In the 20–volt range, it will read up to 19.99 volts. In the 200–volt range, it will read up to 199.9 volts. Finally, in the 1,000–volt range, it will read up to 999 volts. Moving the decimal point will affect the precision of the reading. A **nominal** 12–volt value could be read as 12.14 volts, 12.1 volts, or 12 volts, depending on the range selected. Other meters may scale at 4 volts, 40 volts, 400 volts, and 1,000 volts. In most cases, the DMM range is the maximum voltage minus one, which can be read in that range.

On the other hand, VOMs generally express their ranges as multiples of 10. For most VOMs, the lowest resistance range is designated as X1, meaning that the number value on the ohm scale is multiplied by 1. If the needle is resting on 3, the value is read as 3 ohms (Ω). If the selector is set on X10, the value is 30 ohms for the same needle position. Similar values are indicated for each of the range multiplier designations. The designation K is used on both VOMs and DMMs for 1,000, and M is used for one million.

The term *resolution* is used to indicate the degree of precision attainable by a meter. In common reference to a digital display, resolution refers to the number of places the value can be read to the right of the decimal. For a given meter, the resolution will usually be dependent on the range setting. A typical high-quality DMM may show range resolution for volts as 500.0 millivolts (mV) (0.5000 volts), 5.000 volts, 50.00 volts, 500.0 volts, and 1,000 volts. This means that this particular meter will indicate hundredths of a volt in the 50–volt range but will only indicate to the nearest volt in the 1,000–volt range.

TRUE RMS METERS

These are the top-of-the-line meters in any manufacturer's inventory. However, before introducing the function of a true RMS meter, it will help you to understand the difference between the direct current and alternating current voltage readings on a standard VOM or DMM. The direct current voltage scale represents the potential difference between two unchanging measurement points. If a 1.5–volt battery is being measured, the voltage difference between the two terminals is shown as the direct current voltage on the meter scale. This is not the case with AC voltage. The standard alternating current voltage available for most consumer and industrial use in the United States and Canada is a 60-cycle **sine wave**. The power is a 50-cycle sine wave in most European, African, and Asian countries. However, the sine wave passes through two zero points on each cycle. Sixty-cycle current has 120 zero points and 120 full-voltage peaks each second. Therefore, a voltage reading of 120 volts AC is an **effective voltage** of 120 volts, not a peak of 120 volts. AC effective voltage is defined as *the value of alternating current or voltage that will produce the same heating effect as would be produced by an equal value of direct current or voltage.* For a sine wave, the effective voltage is equal to 0.707 times the peak value. This is the **RMS, or root-mean-square**, value. Or, stated in peak voltage values, a 60-cycle sine wave must reach a peak voltage of approximately 170 volts to equal the heating effect of a 120–volt direct current voltage. Since alternating current voltage is defined as an effective voltage value, meters are scaled to read this equivalent value. The actual voltage is alternating between zero and 170 volts. For convenience, however, your meter will represent that value as a constant 120 alternating current voltage. Or, if you are reading 240 volts, the values are alternating between zero and approximately 340 volts.

The Effect of Non-Sinusoidal Waveforms

Harmonics. For now, a simple example will be used to show the difference between a true **sinusoidal wave**, which can be measured accurately by any standard VOM or DMM, and a non-sinusoidal wave, which can be accurately measured only by a true RMS meter. In Figure 5-10 you see the waveform produced by a household light dimmer. The true RMS voltage is 100.2 volts in this example, whereas a standard DMM indicates 82 volts. This variation occurs because the standard DMM averages this waveform as though it is a true sinusoidal wave, whereas the true RMS value is 100.2 volts, which accounts for the maximum voltage spike of 248 volts in the non-sinusoidal wave. However, the light bulb responds as though it is powered by 82 volts. On the other hand, when the two meters that were used for this test are compared on a household receptacle with clean power, they both show the same value.

Troubleshooting with True RMS Meters

There is no need to know true RMS values while doing the on-line troubleshooting that has been explained to this point in the book. These meters have been introduced here simply to acquaint you with them. These true RMS meters are invaluable in many types of troubleshooting where the electrician encounters non-sinusoidal waves, electrical noise, and harmonics. The electrician will frequently encounter this type of testing in both direct current and alternating current motor drives. True RMS clamp-on ammeters are equally important, as will be seen in Chapter 9.

CHOOSING AN ON-LINE CONTACT METER

On-line resistance testing places the meter's leads in physical contact with the live circuit under test. Consequently, this type of testing will require specialized test instruments that can safely be used on live circuits while they are in the resistance and continuity settings.

In Chapter 4 you were shown how a meter could be used for on-line troubleshooting. In this chapter, you will be shown how to select appropriate meters. Before going further, look at Table 5-1, examining each of the headings at the top of the table.

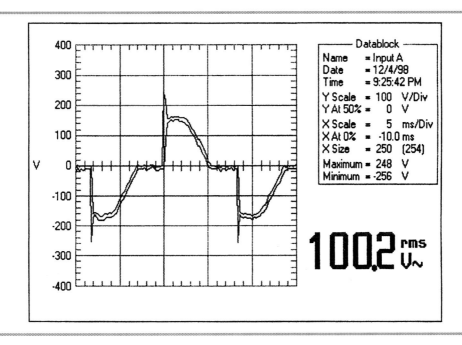

Figure 5-10 A non-sinusoidal waveform does not produce a true value on an averaging voltmeter. This waveform measured 100.2 volts on a true root-mean-square (RMS) meter, but only 82 volts on an averaging digital metre. Notice the peak value of of 248 volts.

On-Line Troubleshooting Tools

Table 5-1 On-Line Troubleshooting Contact Meters

Meters that are suggested for on-line troubleshooting. These meters can be used in a continuity or ohm value mode on an energized system. Some of these meters will automatically toggle between resistance and voltage values.

Meter	Voltage range	Ω range (Volt allow)	Protection	Impedance	Choice for on-line troubleshooting*
AUTO SELECTING METER					
High Resolution DMM					
FLUKE 12B	0–600 AC/DC	0–40M†	CAT III	Low (0.10 mA)	A+/B
FLUKE 16	0–600 AC/DC	0–40M†	CAT III	Low (0.10 mA)	A+/B
FLUKE 7-300	0–300 AC/DC	0–400†	CAT III	Low (0.10 mA)	A/B
FLUKE 7-600	0–600 AC/DC	0–400†	CAT III	Low (0.10 mA)	A/B
Partial High Resolution					
Greenlee 6710	700 AC/DC	50K†		Ø (0.07 mA)	B+[1,2]/A
Low Resolution Lamps					
FLUKE T2	Max 600 AC, Max 220 DC	2000†	CAT III	Ø (0.00 mA)	C++/A
Greenlee 6706	600 AC/DC	800K†		Ø (0.07 mA)	C+[2]/A
Greenlee 6708	440 AC/DC	800K†		Ø (0.07 mA)	C+[2]/A
PROTECTED OHM DMM					
AWS DM-6593A	0–500 AC/DC	0–20M (250 ac/dc)	CAT II	Low (0.13 mA)	B+/A
FLUKE T5-600	0–600 AC/DC Lamp in Ω range	0–1000†	CAT III	Ø (0.09 mA)	B++/A
FLUKE T5-1000	0–1000 AC/DC Lamp in Ω range	0–1000†	CAT III	Ø (0.09 mA)	B++/A

Nomenclature: * A (highest), B (moderate), and C (lowest) preference for meter display values. /A (highest), B (moderate), and C (lowest) preference for meter comfort while in use.

† Meters which will allow full rated voltage while in the ohm range.

Notes: [1] The Greenlee 6710 would be rated higher except that continuity is identified with an audible signal only. It does not use an LED to signal continuity as do the Greenlee 6706 and 6708. [2] These meters are polarity sensitive. The continuity LED and/or audio will squawk when the meter probe is placed on a hot terminal without the lead in contact. This can be eliminated by reversing the polarity of the meter probe and lead while testing. Because of this feature, care must be taken to avoid confusion between continuity and a hot open circuit even though the tone is dissimilar between the two conditions.

There are seven criteria that can be used for meter selection. Five of these criteria are given in Table 5–1.

1. *Voltage range.* The voltage range identifies the maximum value of either alternating current or direct current that the meter can measure. When the meter can display sequential numeric values, it is shown as a 0-maximum value. The term *Lamp in Ω range* indicates that the meter does not display a numeric voltage value when the meter encounters a voltage in the resistance setting.

2. *Ohm range.* The ohm range identifies the maximum resistance on which a reading may register. If the meter is capable of sequential numeric display, the value will be shown as 0-maximum. In the case of low-resolution lamp meters, the lower maximum value is preferable for on-line troubleshooting. An open circuit is more likely to indicate continuity on a meter with a maximum 800K range because the meter can read through other parts of the circuit. A meter with only 50K of range will be more selective to the actual circuit under test. However, many of these meters will produce a discernibly different audio signal at low resistance than at high resistance. *Volt allow* in brackets indicates the maximum allowable voltage tolerated by protected-ohm meters when they are in the resistance range. Generally speaking, the designation "overload

protection" does not identify an instrument capable of doing continuity testing on a live circuit. A designation indicating a maximum voltage (usually 250 volts) for the resistance and continuity modes may allow on-line continuity testing. It will generally show the symbols for continuity and ohms followed by "250 volt max." You must carefully read the operator's manual to determine exactly what voltage limits the manufacturer is specifying. For obvious reasons, any auto selection meter will allow full voltage while in the ohm range. These meters have been identified with a dagger (†) in the ohm range column.

3. *Protection* indicates the level of protection warranted by the manufacturer. Any meter that has a protection rating as CAT II or CAT III is protecting the user up to that specified overvoltage in *all* meter functions, including resistance and current. Of course, this does not mean that the meter itself will not be damaged by this degree of overvoltage. This category must be weighed heavily in the selection of any hand-held meter that is used around line voltages, which often exceed 480 volts.

4. *Impedance.* This category identifies the meter's impedance in its ohm setting. The internal circuit of a meter, which can be used on an energized circuit in its resistance (ohm) range, has considerably lower impedance than a DMM in a voltage range and will draw higher current when placed across a voltage. Though the meters identified are designed for this application, the meter's low impedance may close small relays or other output devices if it completes an open circuit. For this test, each meter was placed in series with an AC milliammeter across an open circuit feeding a 120-volt AC ice cube relay. If the meter caused the relay to close, it was identified as *low* in the *Impedance* column. If the meter did *not* cause the relay to close, it was identified with a Ø. After the circuit stabilized, the current draw of the meter was recorded in parenthesis. The relay used for this test draws 0.17 milliampere on 120 volts, though it will close with as little as 0.09 milliampere when there is voltage loss through a meter. However, a *low* rating indicates that the meter must be used with particular care for on-line troubleshooting in its resistance (continuity) range because small relays or output devices could be cycled by the test itself. Meters that provide a function selection between volts and ohms would have high impedance in the volt setting and would not pose this potential risk. Larger relays or motor controllers should not be affected by the current draw of these meters. *Nonetheless, the effect of any meter on the circuit being tested should be verified before attempting on-line troubleshooting.*

5. *Choice for on-line troubleshooting* indicates my personal assessment of an individual meter's use for on-line troubleshooting as defined in Chapter 4. You must bear in mind that this is my subjective evaluation and it considers only on-line troubleshooting. Two ratings are given that grade the meter from *A* (highest), *B* (moderate), to *C* (lowest). The first rating considers only the meter's display. High-resolution digital readout displays in absolute numbers are given an *A* grade. LED on-off status lights are given a *C* grade. A combination of digital readout and status lights is given a *B* grade. Meter comfort is also graded from *A* to *C*. The general criteria considers the shape and probe construction of a meter. A "box" meter with two probes is hard to control in a panel when two hands must be used for the probes and the meter itself requires a "third hand." This configuration is given a *C* grade. The most convenient meters to use in panel troubleshooting are those with a hand-grip meter body, which includes either a permanent or a demountable probe on the meter itself, and a single extension probe for the second hand. This configuration is graded *A*. A "box" meter with a probe clip is rated in between as *B*. Again, however, the comment must be made that what is most comfortable to use in panel testing may not be the best selection for other types of testing.

6. *Distinctive audio tones.* Meters that use audible signals will sometimes use a clear tone for full continuity, and a periodic beep or a squawk for higher-resistance conductivity. This distinction is valuable because the circuit under test may be open even though there is feedback from other parts of the circuit. The best way to evaluate this function *when purchasing* a meter is to carry a fixed resistor of 100, 500, and 1,000 ohms in

your pocket. If you cannot tell the difference between direct probe contact and contact through at least one of these resistors, you will have trouble discerning the difference between full continuity and a high resistance contact in a panel.

7. *Other uses for the meter.* Category number 5 considers the meter's use for only the on-line troubleshooting task. Generally, you will use the meter for other testing besides malfunctioning relay logic circuits in an electrical panel. You will need to consider these other uses in the final selection of a meter. For my personal use, if I could carry only one meter, it would be the FLUKE T5-1000 because it carries a CAT III DMM with up to 1,000–ohm range with the additional bonus of an ammeter. My second choice would be either a Greenlee 6710 because it provides a digital voltage readout and is an extremely convenient instrument to use or the FLUKE 12B. My third choice would be the pencil meter by AWS. This is a versatile and compact meter, and has the advantage of full DMM functions. The CAT II rating of the AWS instrument ensures greater personnel protection from electrical hazard. Either of the Greenlee 6706 or 6708 are also extremely useful test instruments to carry in a toolbox.

8. *Price.* The final, but necessary consideration, is price. However, all of these meters were selected for this evaluation because they are in the budget range of any journeyman electrician. High-priced, specialty instruments were avoided.

The ideal meter for relay circuit on-line troubleshooting would be an auto-selecting, high-resolution DMM, which would display full-voltage and resistance ranges. The meter would carry a minimum CAT III safety designation, and would be convenient for two-handed use. Since no single meter supplies all of these features, each electrician much choose which meter best suits his or her needs, considering the features that must be compromised in light of other testing for which the meter will be used.

All meters listed in Table 5–1 can ostensibly be used for resistance testing on an energized control circuit of 120 volts or less. In all cases, however, it is the electrician's responsibility to read the material supplied with the instrument, verifying that the meter is appropriate for the intended application and is being used safely. This list is only representative. There may be other meters capable of performing these same tests.

You were given a complete description of the use of contact meters in the troubleshooting examples of Chapter 4. That information will not be repeated in this chapter.

NON-CONTACT ON-LINE TROUBLESHOOTING

There are a number of non-contact—or proximity testers—on the market today. They are useful for locating energized lines, checking fuses, testing a component by checking the "in" and "out" leads, and many other tests where you want to determine if an insulated conductor is live. However, when you are doing on-line troubleshooting inside a crowded electrical panel, you must be aware of the instruments' limitations. Because these instruments are sensitive to the surrounding electrical field, the density of the field in a control panel makes it difficult, if not impossible, to isolate a single wire for testing. Generally, it is not possible to discriminate between wires closer than one-half inch apart. With many wires in the same vicinity, the distance grows to greater than one inch. However, an instrument with a sensitivity adjustment may be more selective. In spite of these limitations, proximity testers can be used for on-line troubleshooting. Refer to Figure 5–11 for examples of various manufacturers' proximity meters.

You are already familiar with the diagram for the mold-close solenoid (solenoid F) of line 19. As you already know, any malfunction that leaves the circuit open will prevent the solenoid from cycling. In the on-line troubleshooting technique in Chapter 4, you set the machine so that it would stall when it came to the inoperative function. Now, referring to Figure 5–12, it is obvious that if relay contact cr7 were to fail, the mold solenoid would not cycle. Consequently, the purpose of troubleshooting is to find the last point on the conductor carrying the control voltage. That point will be adjacent to the electrical device that is inoperative or open. Because you were told in the last example that the problem is with cr7, the last point on the conductor that will carry the control voltage will be the terminal screw of cr7 for wire 38.

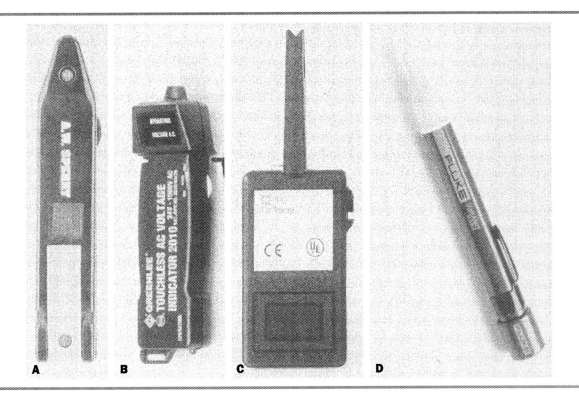

Figure 5-11   A sampling of four manufactures' proximity testers. (A) The Volt Sensor proximity tester by A. W. Sperry, and (B) the Touchless AC Voltage Indicator by Greenlee both have variable sensitivity control. (C) The Tic Tracer by Tif has a long probe that is useful while reaching into congested areas. (D) The VoltAlert by FLUKE can be conveniently carried in a shirt pocket. *Courtesy of A. W. Sperry Instruments, Inc., Greenlee Textron, Tif Instruments, Inc., and FLUKE Corporation.*

Every troubleshooting technique using the ladder diagram to find open circuits essentially uses some form of electrical information to locate the final point of conductivity on that particular control wire. Consequently, voltages are measured from various points, or continuity is read to locate open circuits. The information is then used to define the location of the open circuit. A non-contact proximity tester reads that information directly from the outside jacket of the conductor.

Actual Testing

You are now going to troubleshoot line 19 from Figure 5–12. You have already determined that solenoid F is not functioning. You have stalled the machine so that solenoid F should close and everything has been left in the operational mode. You are now ready for the test. For this example, you will use an AWS Volt Sensor (Model VH-601A) proximity tester because you can adjust the sensitivity to eliminate false readings in the panel.

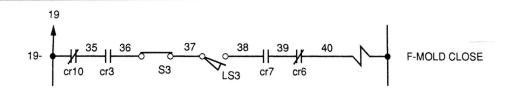

Figure 5-12   Diagram line 19.

72 CHAPTER 5 On-Line Troubleshooting Tools

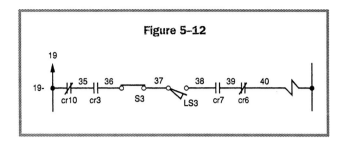

Figure 5-12

Test Point 1
Verify the tester's operation. In order to avoid false readings in the close quarters of an electrical panel, the tester must be in its lowest sensitivity range. Set the sensitivity adjustment by increasing the sensitivity from its minimum setting while holding the tester's tip next to a known energized wire. Select a minimum setting where the tester first responds with a steady tone and an LED signal. Always keep the tester as far from other wires as possible and touch each wire using the same tip surface and angle.

Test Point 2
Arbitrarily select a midpoint on line 19 and test for a voltage. Hold the tip next to wire 38.

Test purpose. To verify the presence of a control voltage.

Test results. A steady tone indicates control voltage. You know the break in the circuit is beyond wire 38.

Test Point 3
Select a point between the solenoid and the last test. Hold the tester's tip next to wire 40.

Test purpose. To verify the presence of a control voltage.

Test results. An absence of tone or LED indicates the absence of a control voltage. You know the open circuit is between wire 38 and wire 40.

Test Point 4
Select a point between wires 38 and 40. Test wire 39.

Test purpose. To verify the presence of a control voltage.

Test results. An absence of tone or LED indicates the absence of a control voltage. You know the open circuit is in cr7.

Test Point 5
Place the tip on both terminal screws of cr7.

Test purpose. To verify cr7 as being faulty.

Test results. A steady tone and LED on wire 38 of cr7 and an absence of tone or LED on wire 39 of cr7. You know the relay contact is open.

It should be apparent that the troubleshooting sequence just demonstrated is extremely fast. It also has the safety advantage of testing conductors without physically touching any part of the circuit with conducting probes or leads. However, care must be taken so that you do not allow stray fields to give you false information.

Note: Non-contact testing never ensures that a particular wire is energized because other closely spaced parallel wiring can induce a field into a deenergized wire. A sensitivity selector may reduce false readings, but it does not guarantee that a meter response indicates an energized wire. On the other hand, if the meter has been verified to respond to energized wires, the absence of a meter response indicating a deenergized wire is the only certain proximity meter response.

The testing pattern for non-contact testing is the same as the one used earlier in the example of on-line troubleshooting given in Chapter 4. In both cases, you divide the circuit and test it at a midpoint. Depending on the results, you subdivide the circuit into smaller sections until you isolate the break in the circuit. The difference, however, is that you are not testing for continuity with this non-contact test. Rather, you are testing for the presence of a voltage on a line, which indicates that the line is unbroken from the source to the point at which the meter is beeping.

There is an advantage with non-contact testing in that you are not using two meter leads as you would with continuity testing. It is simpler to hold the nonconductive probe against a single conductor.

CHAPTER 5 On-Line Troubleshooting Tools 73

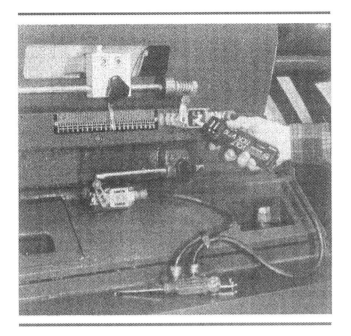

Figure 5-13 A proximity meter being used to test a limit switch circuit. Because the lead to this limit switch is flexible cord, the test could also be done on the cord itself without removing the limit switch cover. However, a meter response from the cord would not identify a problem in the limit switch. A proximity meter could not discriminate between one or two energized wires.

In addition, you do not need to locate an open terminal for the testing since you can take the reading from an insulated wire. This advantage may save you the effort of removing a cover plate on a limit switch. If the switch is closed, both wires leading to the switch will test as energized any place on the wire. If the switch is open, only one of the two wires will test as energized. Note the example in Figure 5–13.

A *caveat emptor* ("buyer beware") clause needs to be added. Panel **EMI (electromagnetic interference)** is high in any electrical panel. In addition to EMI, wire bundles are closely spaced and can interfere with the wire you are attempting to test. With experience, non-contact tests can be successfully used in both electrical panels and on remote areas of industrial machinery. But do not buy expensive proximity testers assuming that they will replace contact testers for troubleshooting work. You will need to carefully experiment to see which meters will perform satisfactorily on the equipment you routinely troubleshoot.

Available Testing Equipment

Table 5–2 lists a number of manufacturer's proximity meters. Again, these meters have been given a grade of A through C for on-line troubleshooting. The basis for this grade is the availability and usefulness of the meter's sensitivity adjustment. Secondarily, the usefulness of the probe in testing among grouped wires in an electrical panel was also considered in assigning this grade. As before, there may be features in proximity testers that are discounted for on-line troubleshooting functions, which make them highly desirable in other applications.

TROUBLESHOOTING WITH SPECIALIZED CONTINUITY TESTERS

There are a number of continuity testers available for deenergized circuit testing. TIF Instrument supplies

Table 5-2 On-Line Troubleshooting Non-Contact Meters

Proximity meters that can be used for on-line troubleshooting. These meters do not require electrical contact with the machine's control circuit.

Manufacturer	Model	Sensitivity adjustment	Audio	LED	Continuity test	Protection	Choice for on-line troubleshooting*
AWS	VH-601A	Yes	Yes	Yes	No	CAT II	A
Fluke	VoltAlert	No	Yes	Yes	No	CAT III	C
Greenlee	2010H	Yes	Yes	Yes	No		A
Greenlee	1010	No	Yes	Yes	No		C
TIF	Tic Tracer 100	No	Yes	No	No		B
TIF	Tic Tracer 300	No	Yes	No	Yes		B+

Nomenclature: *A (highest), B (moderate), and C (lowest) preference for meter response in use.

two such testers. Their Tic Tracer 300 has two leads that are used for continuity testing, in addition to the proximity tester already described. TIF also supplies a capacitor tester (model TIF660), which has a continuity function. Both of these meters are shown in Figure 5–2. Other suppliers have similar testers. Most use probes or a combination of probes and alligator clip test leads.

These testers generally indicate continuity with an audible signal. They are used exactly like a VOM on the deenergized circuit resistance tests. The only difference is in the meter's audible signal response rather than an analog or digital display movement.

The choice of an audible continuity tester response over a multimeter for deenergized circuit continuity tests is largely a matter of preference and the background noise levels in the plant itself. In some cases, however, there can be distinct advantages with the audible response. It is often easier to listen for a signal than it is to look at the indication portion of a meter while you are attempting to use the lead probe on wire terminals. Many times these audible testers are more compact, and even though they lack greater application, they are generally less expensive.

The same precaution in testing only deenergized circuits applies to these meters as it does to VOMs and most DMMs. Unless specifically stated that the given tester is designed for live circuit testing, the circuit must be deenergized before any testing is done.

TESTING PROGRAMMABLE LOGIC CONTROLLER CIRCUITS

New equipment is increasingly using programmable logic controllers (PLCs) rather than relay logic circuits. In many cases, the external control switches, limit switches, and other make-or-break parts of the control circuit will be similar to conventional relay-controlled circuits. However, you must practice extreme caution in testing these circuits when the PLC is energized.

A voltage applied across an open contact may energize the circuit of a PLC. Even the lowest current producing solid-state meters may introduce enough current into the control circuit to energize an **input port** of a programmable controller. Therefore, if you are testing a circuit that is conductive to the input side of the PLC, you run the risk of cycling a controlled portion of the equipment. In other parts of the book, you learned the need to set equipment so that it can safely cycle when you are doing on-line troubleshooting. With a PLC, however, you may have an entirely different condition. You can actually *energize* an otherwise open circuit with your meter. Indiscriminate testing poses the risk of doing great equipment damage to say nothing of the risk to personnel.

Therefore, unless you completely understand the circuits you are testing, do not try to use on-line troubleshooting techniques with PLC-operated equipment. Nonetheless, input devices such as limit switches and selector switches can be tested with a continuity meter when the PLC is secured in a deenergized state.

However, most external input device troubleshooting for PLC applications will use the LED readouts on the PLC itself. When an input circuit is *true* (complete), the LED for that input will glow. Consequently, troubleshooting the input devices will comprise reading the prints to determine input numbers and then verifying that the appropriate LEDs light.

For example, if you have a conveyor that is PLC controlled but is not moving forward, you would verify on the printed program which inputs must be true for the chain to advance. If the conveyor is powered up but not advancing, looking at the PLC's LEDs will indicate which inputs have properly cycled. This particular chain may require that a loading area proximity sensor be blocked, an unloading area proximity sensor be unblocked, and a momentary push button at an operator's station has been pushed. You may physically examine the conveyor line to make certain that a box is on the loading area and the unloading area is open. Then you would ask the operator to hold the push button in the maintained position while you look at the input LEDs on the PLC. Say you discover that the push button LED is glowing, the unloading area proximity sensor LED is not glowing, and the loading area proximity sensor LED is not glowing. You will expect to find a problem with the loading area proximity sensor. A careful physical check shows that the sensor has been bumped and moved out of range. Assuming that there has been no damage to the proximity sensor, readjusting it will solve the problem.

This is certainly a troubleshooting procedure. But it uses the *PLC's self-diagnostic LEDs* rather than electronic meters for the testing.

CHAPTER REVIEW

Advances in electronic test equipment technology have brought significant improvements in the speed and accuracy now possible to the maintenance electrician's troubleshooting procedures. The reduction of troubleshooting time will mean an increase in machine productivity. In order to achieve the full potential of the present electronic testing equipment, however, the procedures and methods used by the troubleshooting electrician need to be updated commensurate with the capabilities of the equipment that is now available.

Electrical testing is generally done with either contact measurement or non-contact measurement techniques. In contact measurement, the leads of the test instrument are in physical contact with the live circuit under test. That is, the circuit of the test instrument is common with the live circuit. In non-contact measurement, the test instrument is sensitive to the electrical field of the conductor being tested, and the circuit of the meter is isolated from the conductor it is testing.

Until recently, the analog moving coil has been the standard meter movement in most electrical measurement equipment. The moving-coil meter has a spiral spring-biased armature placed between two pole magnets. A properly calibrated coil will rotate linearly in response to a linear voltage applied to the armature terminals. The full voltage range of a moving-coil meters coil is generally less than 0.05 volt. All voltage functions (alternating or direct current), resistance measurements, and current measurements can be read on this meter when the appropriate resistance, rectification, and battery-powered circuits are added.

The volt-ohmmeter (VOM) and digital multimeter (DMM) are used in similar fashion in electrical troubleshooting. These meters are used for: (1) voltage value tests where an actual voltage value is sought, (2) voltage status tests where the mere presence or absence of a voltage is sought, (3) live circuit continuity tests where the voltage ranges are used to indicate an unbroken conductor, and (4) deenergized circuit resistance tests where the resistance ranges are used to measure continuity of deenergized conductors.

Because of recent advancements in the field of electronic testing equipment, two new testing procedures are now available to maintenance electrical troubleshooters. True continuity testing can now be done on live circuits, and field-sensitive, non-contact testing equipment is available that allows the tracing of an energized conductor requiring no electrical contact.

THINKING THROUGH THE TEXT

1. Define and explain contact testing.

2. Define and explain non-contact testing.

3. In what ways is an analog moving-coil meter similar to a permanent magnet direct current motor?

4. A voltage meter will show an alternating current or direct current voltage as an effective voltage value. How is the comparison value of alternating current and direct current voltage derived?

5. What precautions are given when using a VOM for voltage or resistance testing?

6. Name and briefly describe the four types of tests for which a VOM is most frequently used in plant maintenance diagnosis.

7. What do the X1, X10, and X100 designations mean on a multimeter's resistance ranges?

8. What does "overload protected" mean on a multimeter?

9. Every troubleshooting technique using the ladder diagram essentially uses test information to locate the final point of conductivity. What test information is generally used? How is it obtained in the circuit?

10. What does the term *resolution* mean when referring to a digital multimeter display?

CHAPTER 6

Collecting Information

OBJECTIVES

After completing this chapter, you should be able to:
- **Gather general information regarding the site and the equipment you will troubleshoot.**
- **Question an operator to determine how the machine normally functions and what took place prior to the machine failure.**
- **Obtain information that will help you isolate the electrical malfunction.**
- **Learn how to use information from the failure itself to prevent future problems.**

STOP, LOOK, AND LISTEN

No better advice could be given for the first step in electrical troubleshooting. *A significant percentage of the equipment failures are not electrical problems in the panel—they are something outside the panel which can often be seen with careful observation.*

Inspect the Physical Equipment

Each type of equipment has its own set of checkpoints. Nonetheless, as you begin the process of troubleshooting, make an inspection of the physical equipment your first step unless there are clear indications otherwise. If the motor is dead and the motor contactor is smoking, you will be forgiven for checking that first rather than inspecting a safety gate that is ajar! Of course, as you get to know the equipment, you will bypass many checks. If the motor is running and the control panel indicator lights are glowing, you know that the main power and the control power are operational—your first checks will not be in these areas. But get into the habit of looking knowledgeably at the physical equipment before you assume that the problem is caused by an electrical malfunction.

Sharpening your sense of observation will result in more than troubleshooting time savings after an actual failure. It is also an invaluable preventive maintenance and safety skill. Many failures—electrical, mechanical, structural, etc.—give ample early warning signs to the observant maintenance person. Excessive heat or flashing around contacts, high noise levels, hot conduits or boxes, and many other sensory perceptions may be an indication that work is needed. If the warning signs are caught early enough, there may be significant savings in reduced equipment losses.

> **A DAY AT THE PLANT**
>
> *An electrical contractor installed a large walk-in washing machine for tray racks in a cookie bakery. During start-up testing, the washer would begin filling and then immediately shift to the wash cycle. The water sump would not fill. The contractor assumed that the fill controls were at fault, and spent time troubleshooting without success. When I came, it would again start to fill, then immediately switch to its wash cycle. I started with an evaluation of the machine. It had a sump under the floor level with a "spark plug" water level sensor. The factory-installed fill pipe was on the same side of the cabinet as the water sensor. It had a nipple in the cabinet wall, a 90° elbow on the nipple, and a 2-foot discharge pipe pointing toward the sump. I had a hunch! We started the fill cycle again. Interlocks required that the walk-through door be sealed before the washer would start. I waited with my hand on the door handle. As soon as the fill cycle stopped, I quickly opened the door and looked at the water level electrodes. Water from the vertical fill pipe was splashing against an internal frame member, running across the frame, and dripping on the water level electrode. The "electrical" repair was nothing more than rotating the fill pipe at the elbow. The washer then filled and operated normally. I dried off and left!*

Good Observation Enhances Safety

Safety is another area in which keen observation may be highly rewarded. Electricians are frequently around machinery and equipment that has high potential for injury to either personnel or the equipment itself. That is certainly true of the electrical part of the installation, and is usually the case with the mechanical and structural elements as well. Again, by being alert to abnormalities, you may well spot something that can be corrected before it causes expensive damage.

Inasmuch as safety and observation are the topic, take this a step further. It seems as though the individuals who are the "shop hazards" are more than just reckless. In many cases, these individuals lack the ability to conceptualize how something could be dangerous. In some cases, their propensity to precipitate accidents may be a simple lack of knowledge. In other cases, however, it is clearly a lack of thinking through the consequences of what they are about to do next. In contrast, *the individuals who work safely do so because they have a sense of what might happen later if their present action is done incorrectly.*

True safety practice comes from the ability to visualize hazardous conditions *before* they occur. The truly safety conscious person is the one who is able to modify or prevent conditions before injury or damage takes place.

TROUBLESHOOTING SEQUENCE

Unless you have specific information that will more quickly isolate the problem area, follow this type of sequence in most electrical troubleshooting problems:

1. *Get information from the operator(s) and supervisor(s).* This is an art! You will soon learn that if the machine was at fault, the answer to what happened will be, "I don't know, it just quit." On the other hand, if the operator was at fault, then the answer to the same question will be, "I don't know, it just quit." Notice the similarity of the two answers! So, you will need to refine your questioning skills, taking into allowance both the machine and the operation. You will want to ask questions such as, "How was it running before it malfunctioned?" "Were there any unusual sounds, smells, or machine sequences when it was last running or when it quit?" "What positions were the switches in?" "What switches have been changed?" "Did you free any jammed parts or change anything?" "Was anything dropped, unplugged, or *bumped*?" Finally, if safety allows,

A DAY AT THE PLANT

There is a story worth telling because it illustrates an employee who used excellent judgment in anticipating a hazard. I worked in a plant that leak-tested large diameter welded pipe with high pressure water. The testing was done in a hydro-test machine, which sealed the open ends of the pipe between two 6–foot diameter gasketed steel plates. The multiple-ton clamping force was provided by a large pneumatic cylinder connected by hoses to the plant air system. An operator violated procedure and reached between the clamp and pipe to adjust a gasket. He had also left the manual valve ajar so that the clamp was creeping closed as he was struggling with the gasket. His torso became wedged between the slowly closing clamp and the edge of the pipe before he started yelling for help. A welder in the area responded, but was uncertain which way to move the valve to open the clamp. If he had turned the valve the wrong direction, the operator would have been killed. The welder was carrying a knife and quickly cut the air supply line.

I can think of no better example of someone who visualized a hazardous condition and acted accordingly. The operator suffered no serious injury because the welder had presence of mind enough to cut the hose rather than trying to open the clamp with the manual valve.

Figure 6-1. Before actually testing circuits, a careful troubleshooting electrician obtains information from the machine operator.

you might want to ask the operator to cycle the equipment again. If you watch carefully, you may discover an operator-related cause. In Figure 6–1 the electrician is verifying the switch position when a malfunction occurred.

You will need to evaluate the information you receive carefully. Not everything the operator tells you is necessarily correct. The real problem may be that the operator assumed that a switch was in a certain position, or assumed that the machine had completed a sequence, and the like. Even so, do not make the mistake of discounting what the operator has to say. He or she has spent considerable time running that piece of equipment. The operator may not understand the technical reasons behind the machine's operation, but nonetheless, you will find that a good operator has an extremely well-developed awareness of when something is malfunctioning on the equipment. So carefully think through what you have been told. Many times the operator's description will give you the necessary clues for quickly finding the problem.

In a larger plant, the maintenance personnel may be directed to get repair information from department heads. In some instances, this will be an advantage because the department head knows both the equipment and the operators, and his or her candid answer will get you closer to the real problem. On the other hand, the reverse can also take place. At times, the department head does not really know what happened, or does not want the maintenance department to know. At this point, a "standard description" of the equipment fault may waste a great deal of troubleshooting time.

If you are not familiar with a particular machine, you will need to have the operator explain how to run the equipment. You may need to have an operator or setup person explain various equipment functions and settings to you. Always ask for help. There is no need for the electrician to feel that he or she is above learning from operators. Asking the right questions will be the fastest and safest way to get the equipment back on line. Your speed in getting equipment into production will be the basis on which you will be judged as a maintenance electrician.

2. *Examine the equipment.* Unless you have good reason for suspecting a specific area, your first troubleshooting task is to make an overall inspection of the entire machine. Depending on the type of installation, this may also include an evaluation of the plant's electrical system, starting with the service panel. You will want to look at anything that has limit switches, electrical safety switches or interlocks, proximity or sensor switches, and the like. Look carefully at those types of switches that are adjustable or subject to abuse, such as door switches and conveyor limit switches. Very often these switches will become loose and move out of operational range, or will have become damaged. It will surprise you how many times you will find a limit switch on a door or cam that will "click" if you move the arm after the door or cam is supposedly in the activated position. You will also be looking for adjustment problems, jammed parts that will prevent electrical or mechanical functioning of the equipment, tripped circuit-breakers or overloads, damaged cords or conduit, and anything else on that specific piece of equipment that would affect the electrical function. You will also want to look carefully at panel control settings, such as hand switches and timer settings. Incorrect settings can shut machinery down just as effectively as electrical panel problems.

A DAY AT THE PLANT

I worked in a plant that used clock-controlled boilers to cure cement overnight. On one particular morning, both a mechanical maintenance man and I spent several hours trying to determine why the boilers had not fired the night before. Our conclusion was simply that the foreman had failed to set a switch on a remote station. By getting the maintenance crew involved, the foreman shifted responsibility from himself to the equipment. If you become a troubleshooting electrician in a large plant, you would do well to establish rapport with operators. Informal communication may result in much-needed troubleshooting information.

Another lesson comes from the same plant. I worked under a maintenance foreman who had a great ability to find out what really caused equipment failures. He was easy-going on people in general, and never openly lost his temper or blamed operators. Many times, our maintenance crew (myself included) would survey a breakdown without knowing exactly what (or who) had caused it. Then some time later our foreman would admit that he knew what had happened, though he would not name individuals. He and I were talking after his retirement luncheon, and he told me his secret. He had discovered that if he didn't blame people, didn't get angry, and didn't name them as the one responsible, he was often told the complete story when he was talking with people privately. It made him a better foreman and a more effective troubleshooter.

> ### A DAY AT THE PLANT
>
> On one assignment in my early troubleshooting days, I learned why listening to operators is important. I was asked to troubleshoot a die-casting machine. An electrical company had spent the day on it and failed. When I arrived that evening, I desperately wanted to succeed! After much effort, I also failed—though this happened before I was using on-line troubleshooting. When I first arrived, the die-cast operator told me that a ladle of molten aluminum had been dumped on an exposed conduit. If I had been listening, I would have realized that he told me everything I needed to know. The electrician who followed me found a conduit full of shorted wires with melted insulation!

3. *Do preliminary power checks inside the panel.* Your first check after starting work inside the panel will be on the main power supplies. Depending on what functions you have lost, you will check the load side of the appropriate circuit-breakers and/or fuses. You will then want to make certain that the control transformer is supplying power to the control circuit.

4. *Study the ladder diagram.* Your first step with the ladder diagram will be to isolate the functions that are inoperative. This will take you to some output device or component on the right-hand side of the diagram. However, before you actually begin your electrical testing, think through the significance of each device on the line in question. A limit or function switch on the ladder diagram may be a clue to something that you need to check on the machine. You may also find something on the diagram that raised questions as you were doing your preliminary check. You may find a timer, a disconnect circuit, an optional function switch, or any number of things that need further confirmation. *Studying the ladder diagram can often solve the problem with no electrical testing necessary by showing you what must be operational in order for that circuit to function.* If the ladder diagram shows a safety limit switch on a door that had just been opened for cleaning, this may tell you all you need to know to put the equipment back into operation. Find the limit switch and listen carefully while you move the control arm. If the door did not fully close the switch and you hear it click after further movement, there is a high probability that you eliminated the problem. At this point it merits attempting to re-start the machine before further testing.

5. *Do the electrical testing.* At this point, you are ready to do the troubleshooting as described in this book.

6. *Complete the post-repair reassembly.* This sixth step needs to be added because it is so often compromised. There is often pressure on maintenance personnel doing troubleshooting work to get equipment back into production. But that does not allow the plant electrician to jeopardize future reliability of the equipment or to create electrical hazards in the workplace. The job must be completed. Appropriate repairs must be made. Faulty equipment must be changed. Covers must be replaced, and the entire job must be completed so that the equipment can be safely operated after the repair.

7. *Complete the post-repair **documentation**.* Finally, documentation must also be updated. If any changes to mechanical or electrical equipment were made, they must be noted on the appropriate shop drawings. This is particularly true if any alterations were made to the machine's electrical wiring.

Discretion may sometimes be used when doing temporary work in order to complete a production run. But provision must be made to properly finish the job at the earliest possible time. Do not forget, and do not let production schedules prevent it!

A related problem is that of repair expense. If you are working in a plant on a tight budget, the pressure is always on you to spend as little as possible in getting equipment back into operation. The wise troubleshooting electrician will want to establish a reputation of reducing unnecessary costs. On the other hand, makeshift work will undoubtedly cost more in future downtime than repairing the equipment with appropriate electrical parts. Do your best to establish a candid relationship with your supervisors that allows you to replace equipment when it is necessary. Build a reputation that assures them that you are knowledgeable enough to make expenditure recommendations that best serve the plant.

A note is in order at this point. The seven steps of the troubleshooting sequence are given in a logical order. However, as you become experienced with on-line troubleshooting and the specific equipment you are servicing, you will be able to take shortcuts. With experience, on-line troubleshooting is fast enough that many times it is both easier and faster to do the testing from the panel rather than verifying everything in advance. If you suspect that something is wrong on the machine, you will certainly want to check it first. If not, however, after a brief preliminary check, you may go directly to the panel and begin isolating sections of the ladder diagram. As you gain experience with troubleshooting, you will develop a pattern that will give you the best speed.

PREVENTIVE INFORMATION

A teaching hospital places high value on pathology. An autopsy does nothing for the deceased patient, but it may have great benefit for future patients. So, too, a good troubleshooting electrician is one who will take the time to figure out why something failed. Actually taking time to do failed-equipment autopsies may prevent future problems.

The Reasons behind Electrical Failures

Contrary to the feeling of the electrician in the example of the first chapter, electric motors do not "just burn out." They burn out for given, and usually preventable, reasons. The motor loss described in Chapter 1 was a result of a faulty bleed valve. Proper evaluation of why the overloads were tripping could have saved the motor, the motor control equipment, and the expense of the downtime while the compressor motor was being replaced. In order to have accomplished that, however, the electrician would have needed the ability to evaluate available information rather than to merely replace parts.

So, why do motors burn out? There are many reasons. But unless you know the specific reason why the motor you are now replacing burned out, the replacement motor runs a high risk of an early failure also. In the example in Chapter 1, the compressor motor burned out because a bleeder valve malfunctioned. With three-phase motors, the most common cause is phase imbalance. A voltage difference of only 3.5 percent between the phases will increase the motor temperature by 25 percent. A 25 percent increase in temperature will greatly reduce the motor's service life. Yet, in spite of the damaging effects of voltage imbalance, how frequently is phase voltage checked—and corrected—when replacing a motor? The failure is usually blamed on mechanical problems.

Not all examples come from motors. Why do contactors or magnetic starters fail? Too often the answer is, "We don't have time to find out—just replace it." Careful examination of failed contactors may indicate unexpected causes of failure. An aluminum shading coil may break because of metal fatigue. A better grade of contactor with a brazed copper shading coil may solve the failure problem. In a similar type of failure the steel lamination on a moving coil may become loose and shift, leaving a large air gap. The coil current will become excessive and heat will destroy the coil insulation.

Understanding Why Equipment Fails

The intent of this brief section dealing with postmortem electrical equipment evaluation is to encourage you as a troubleshooting electrician to develop the habit of thinking through the "why" of failure. It is true that the electrical equipment you deal with has a finite service life. Too often, however, electricians are replacing equipment that has failed well within its expected service life because it has been abused, or because other controllable factors have been ignored.

Try to understand *why* the equipment failed. Use your test equipment to make measurements of the voltages and the current draw. After replacing a motor, verify that the current is within its normal operating range. You may find that the new motor is drawing excessive current because of an unseen mechanical problem that must also be repaired. Tear the old equipment apart before you throw it in the trash can. Finally, read technical literature that will give you the theory of that equipment's operation.

A plant maintenance electrician is not just a part-swapping artist. The electrical trade needs electricians who are able to evaluate electrical equipment and make judgments that will improve both the economy of the plant and the reliability and safety of the equipment. Certainly, an electrician will replace parts. However, if you as an electrician are able to understand the root causes of the failures and see the warning signs of impending failure, your contribution will be immeasurably greater.

Get into the habit of determining why equipment failed. As a result, you will certainly be a better electrician. The immediate benefit will be improved equipment reliability and service life.

CHAPTER REVIEW

In this chapter, careful observation is emphasized as an important part of good troubleshooting practice. Unless specific failure areas are suspected, a complete examination of the physical equipment is the first electrical troubleshooting step. A habit of careful observation becomes a preventive maintenance and safety benefit. Future failures or hazards may often be avoided when careful attention is given to the physical condition of the equipment.

In the absence of an indication of failure in a specific area, the following troubleshooting sequence is suggested:

1. *Get information from the operator(s) and supervisor(s).* Their experience with the machinery they are operating will often provide a shortcut in isolating problem areas. Asking specific questions will help you to more accurately determine the cause of the machine's breakdown.

2. *Examine the equipment.* An overall inspection of the entire piece of equipment is the first on-site troubleshooting technique. Many problems that appear to be malfunctions in the electrical panel will, in fact, be switch or control problems needing correction on the machinery itself.

3. *Perform preliminary power checks inside the panel.* The first check inside the electrical panel will be the main power supplies followed by the control transformer output.

4. *Study the ladder diagram.* The ladder diagram is first used to isolate the inoperative function in the equipment. The diagram is then used to pinpoint potential problem areas with switches or electrical control devices.

5. *Do the electrical testing.* The ladder diagram is finally used to move through a testing procedure of individual devices or potential trouble areas within the inoperative circuit. Information is also collected for preventive purposes. Understanding why equipment has failed may be an invaluable source of information for the proper protection of the newly installed replacement equipment. Determining why equipment has failed may identify other areas of maintenance need.

6. *Complete the post-repair reassembly.* The troubleshooting job is not finished until the machinery is electrically and mechanically secured and all hazards are removed from the workplace.

7. *Complete the post-repair documentation.* Finally, all necessary documentation and shop drawings must be updated.

THINKING THROUGH THE TEXT

1. What kind of equipment failures can be observed from outside of the electrical panel?

2. Can you give an example, either from the text or from your own work experience, of a nonelectrical (mechanical) failure that simulated an electrical failure in the panel?

3. Unless there are clear indications for a specific electrical problem, what should be your first step in electrical troubleshooting?

4. What important information can the operator give that will help you in the troubleshooting process?

5. How might safety be improved through careful observation of plant equipment?

6. List the seven steps of the troubleshooting sequence.

7. How might examining equipment that has failed prevent future problems?

CHAPTER 7

Practical On-Line Troubleshooting

OBJECTIVES

After completing this chapter, you should be able to:
- Collect preliminary information from a machine operator in order to facilitate your electrical testing.
- Understand the concept of troubleshooting a *stalled* machine.
- Effectively troubleshoot using both voltage and continuity values.
- Use non-contact testing procedures as an electrical troubleshooting option.
- Understand and use special setups for testing when a machine will not stall.

Imagine that you are the plant maintenance supervisor for a plastics injection molding company. Among your other qualifications, you are a highly skilled, state-licensed electrician. You have just been paged by the plant production manager because there is a problem with machine #18. The information you are given is no more specific than "The machine is having a problem with the ejection and is stopping during operation." It is 9:17 A.M.

You walk to the tool room to get your tools. You take a straight blade screwdriver, a Phillips screwdriver, and your test meter. Those are all the tools you will probably need. You always carry a pocket notebook, which you can use for notes as you make your preliminary evaluation. You did not pick up a wiring diagram because that is already taped to the inside of the electrical panel.

COLLECTING PRELIMINARY INFORMATION

By 9:18 A.M. you are standing by the machine. Your first step is to get information from the operator. Through careful questioning, you find that the machine has been running this particular job for two and a half shifts, so you conclude that the malfunction is most likely an equipment failure rather than a setup fault. You also discover from talking to the operator that the problem has been developing over a period of several weeks on all jobs run on this machine. The problem has occurred at such infrequent intervals, however, that it was not noticeably affecting production. When the problem did occur, operators would turn the function switch to manual and eject the part with the ejection push button. After manual ejection, the machine would again function automatically. As a result, the problem had not been reported to the shift supervisor. This morning, however, it has malfunctioned three times. The shift supervisor stopped the machine and called you.

Setting the Machine to Stall

After a quick visual survey of the operator's panel, you find that the main control switch has been turned to the manual setting. The machine is stationary, though the motor is running and the panel lights are lit. Thus, you know the main power fuses, circuit-breakers, and control transformer circuits are operational. Because the machine is in the manual setting, you cannot proceed with stalled on-line troubleshooting.

Since there are no apparent risks in running the machine, you ask the operator to put the machine back into production. You then explain to the operator that you want to be called immediately the next time the machine malfunctions. With a careful explanation of the need for doing so, you ask the operator to touch none of the controls or safety gates until you return.

By 9:22 A.M. the machine is back in operation. Your technique so far has resulted in only 5 minutes of lost machine time. Compare that with the electrician who will shut the machine down and make a pretext of "troubleshooting" before even knowing the possible area of the malfunction. That little charade may accomplish nothing and yet cost an initial 30 minutes or more of downtime.

Initial Data Acquisition

Now is the time to gather additional information. Since you have a ladder diagram taped to the inside of the electrical panel, that will be your first stop. You brought the screwdrivers so you could open the main panel. As you study the ladder diagram, you realize you need more information from the operator. The malfunction could have been in either the ejection circuits or in the mold-close circuit. According to the diagram portion of Figure 7–1, if the part could not eject, the machine could not sequence to the mold-

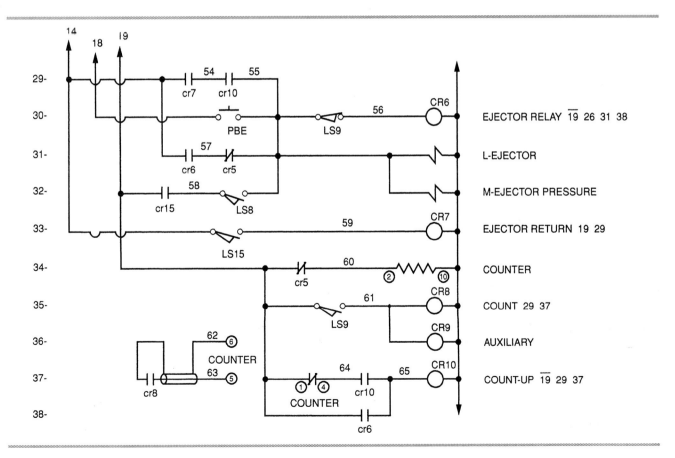

Figure 7-1 Diagram lines 29 through 38.

86 CHAPTER 7 Practical On-Line Troubleshooting

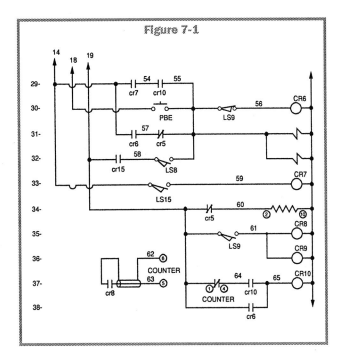

Figure 7-1

close circuit. If the malfunction was in the mold-close circuit, the mold would not close after ejection. Further questioning of the operator gives you the information you need; the machine failed to eject after opening. Now you know that the problem is probably located on lines 29 through 38. More precisely, you know that CR6 (line 30) or solenoid L or M (lines 31 or 32) is probably not sequencing properly.

Your first concern will be CR6 and any of the circuits feeding it from lines 29 through 33. You see that this will also include cr5 on line 31 and cr15 on line 32. CR5 is the mold slow down that prevents the mold from closing at full travel speed, and CR15 is the mold open relay. Though experience tells you that the present problem is probably not a hydraulic solenoid failure, you are aware of solenoids L and M on lines 31 and 32.

While studying the diagram, you notice several things in addition to the relays that should be checked. First, you notice that there are three limit switches in the circuit. There are four limit switch functions because LS9 has two circuits. All are adjustment switches, which are changed during setup. LS8 (line 32) is the ejector forward limit switch; LS9 (line 30) is the ejector stroke; and LS 15 (line 33) is the ejector return. Since these three limit switches are visible through the window on the operator's door, you can visually check them while the machine is in operation to make certain they are not damaged or loose. The limit switches appear to be operating normally; because this malfunction was not a problem encountered immediately after a mold change, you can assume that it is not due to a setup error. You do not eliminate the possibility of erratic limit switch operation, but there is no visual indication of damage or improper setup adjustment.

You also notice on line 29 that count-up relay CR10 is used in the ejector relay (CR6) circuit. Consequently, you know that the counter on lines 34 through 38 could cause CR6 to malfunction.

Before going further, you need to verify the function of the vertical power wires 14, 18, and 19 on the left-hand side of the diagram shown in Figure 7–2. Wire 7 is the main control power. It becomes wire 10 after passing through the master control relay cr1. For safety's sake, no other machine functions are permitted when the mold height is adjusted. Switch S6 is the mold height adjust, which de-energizes control power to all other machine functions when the mold height solenoid is energized

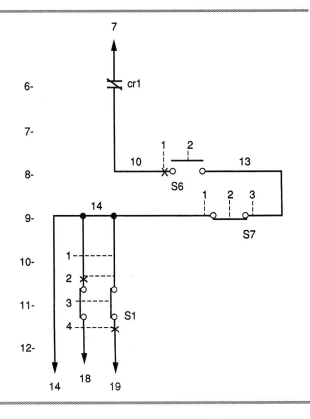

Figure 7-2 The vertical power wires are on the left-hand side of a diagram. Diagram lines 6 through 12.

in switch position 2. Since S6 is in position 1 during machine operation, wire 13 is energized. Switch S7 is a three-position switch with mold height "larger" (switch position 1), mold height "off" (switch position 2), and mold height "smaller" (switch position 3). This redundant circuit ensures that the mold height solenoids cannot be operated when other functions of the machine are energized. Since S7 is in position 2 during normal machine operation, wire 14 is always energized when the machine is running.

Switch S1 is the operation selector switch. All control power is de-energized in "off" (position 1). Position 2 is "manual," which energizes wire 18. Position 3 is "**semi-automatic**," which allows the machine to make a single complete cycle each time the main door is closed. Position 4 is "automatic," which allows the machine to run continuously. Wire 19 is energized in either semi-automatic or the fully automatic operation. Therefore, while the machine is in operation, you know that both wires 14 and 19 are energized.

As you study the circuit, it becomes obvious why the ejector would work in manual even though it would not work in the automatic setting. Line 30 indicates that PBE (push button ejection) bypasses the entire relay-controlled circuit. That is, PBE bypasses relay contacts cr7, cr10, cr6, cr5, and cr15. You have also discovered that wire 18 is energized when the operation switch S1 is placed in manual (position 2). It would be logical that an intermittent problem in the circuit could be reset by cycling the ejector relay. It also gives you an indication that the problem is not in the ejector relay CR6 on line 30 or in either of the solenoids L or M on lines 31 or 32, since a failure in any of these devices or components would not allow the ejector to operate with the ejector push button. Limit switch LS9 on line 30 must also be considered operational for the same reason. PBE bypasses all four circuit components (LS9, CR6, and solenoids L and M), allowing the ejector to work in manual after it has failed in the automatic setting.

PRELIMINARY ON-LINE TROUBLESHOOTING

We generally think of troubleshooting as something done after machine failure. This is not always the case. At times, the preliminary steps for effective troubleshooting can be done on an operational circuit. A faulty circuit can be diagnosed much more quickly after its full function is understood.

Since the machine is back in operation, you can observe the relay action inside the panel as the machine cycles. Even though all three lines (29, 31, and 32) are energized in the automatic mode, you realize that only one would normally be feeding CR6 at a time unless there was an overlap during switching. Careful study of the diagram indicates that lines 29 and 32 are the **initiate circuits** that cause CR6 to cycle. Line 31 is a holding circuit that uses a contact in CR6 to maintain itself until the circuit is broken by cr5. You will verify each of these assumptions with preliminary testing while the machine is running.

You can now do several preliminary tests to verify the function of each of these circuits. For the examples in both this section and the actual on-line troubleshooting section, you will be shown the procedure using both contact and non-contact testing. In the first case, you will be checking for continuity or a possible voltage on a live circuit with a contact tester. In the second case, you will be using a non-contact tester to verify the presence of a voltage.

Your first test using a contact tester will be for voltage. All other tests will be for continuity. Bear in mind that the actual choice between a continuity or voltage test may be arbitrary. In many instances, similar results could be obtained with either. However, in some cases, there will be definite advantages to a continuity test under power, as was noted in Chapter 4. For that reason, this example will use continuity tests unless otherwise necessary.

There is one other qualification that must be made to this example: all tests with non-contact meters are AC voltage field tests made on a single conductor. Because parallel energized alternating current wires in a bundle may induce a field in an open wire, non-contact meters may give a false reading when the circuit is de-energized. The A. W. Sperry Volt Sensor™ VH-601A meter has a sensitivity adjustment that can be used to compensate for induced fields. For the sake of this example, assume that you were able to adjust this meter's sensitivity low enough to eliminate false readings.

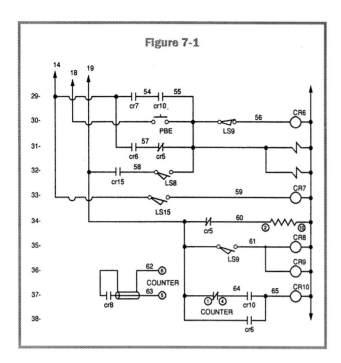

Figure 7-1

Test Point 1

Contact (voltage) test: The meter leads are placed on wire 55 and the common grounded conductor.

Non-contact test: The probe is held next to wire 55.

Test purpose. To establish the presence or absence of a control voltage to CR6 and solenoids L and M.

Meter indication.

Contact (voltage) test: A voltage value is indicated during each ejector cycle.

Non-contact test: A steady tone is heard (or an LED will light) during each ejector cycle, indicating the presence of a control voltage.

Test result. During each ejector cycle, a control voltage is observed for the duration of the ejection sequence.

Note: During normal operation of the machine, the meter can be read—or heard—while observing various relays' actions in the circuit. In this way, you can easily verify the action of line 32 as the "initiate" part of the circuit. Relay contact cr15 will momentarily close at the moment your meter first indicates a control voltage to CR6. However, the voltage will continue after relay contact cr15 opens. Similarly, watching each of the relays in the circuit will show you its function in the ejector cycle. It will be obvious that CR6 (through relay contact cr6) sets the holding circuit, which is broken when cr5 is opened.

Test Point 2

Contact (continuity) test: The meter leads are placed on wire 14 and wire 55.

Non-contact test: The probe is held next to wire 55.

Test purpose. To establish the sequence of the control circuit feeding CR6 and solenoids L and M. A single circuit will be indicated irrespective of which of the three lines (29, 31, or 32) is conducting.

Meter indication.

Contact (continuity) test: A steady tone will sound when the circuit is complete.

Non-contact test: A steady tone will sound (or an LED will light) when the circuit is complete.

Note: Some of the low-resolution auto-function contact meters suggested for this testing are highly sensitive to grounding. If you hold the lead probe against the hot side of the circuit, your hand will offer enough capacitance through the case of the meter unit to give a low-grade tone. Though you can accurately use the meter this way if you allow for the tone variation produced by your own body capacitance, it is less confusing if you hold the tester so that the meter unit's probe is always held against the common grounded conductor side of the circuit, and the lead probe is used to test the unknown wires or contacts. It is always a good procedure to do an initial check on both a known live and a known deenergized wire for comparison before you start the actual testing.

Test result. Under normal cycle conditions, the test will indicate that there is a closed circuit between either wires 14 or 19 and wire 55 during the entire ejector cycle time.

By this time you have done almost all you need to do in the panel itself. However, a last series of tests might be worthwhile to give you a better understanding of each of the contact points in the circuit. This testing could only be done with a high-resolution auto-function DMM or a protected-resistance DMM. By doing a resistance test across each relay contact or limit switch, you can isolate that contact for observation during actual operation under load.

Test Point 3

Contact (resistance) test: The meter leads are placed on any two wires separated by a single contact point, as for instance, wires 54 and 55 for relay contact cr10.

Note: If you are using a meter capable of reading resistance (in ohms) with a voltage on the circuit, you can read the actual resistance value across the closed contacts while they are under load. It is sometimes possible to spot a potential problem in this way since the resistance reading is often abnormally high before complete failure. However, carefully follow the manufacturer's recommendation concerning test duration. These meters have sensitive solid-state circuits, which must dissipate heat. Even though some of these meters can tolerate a voltage on the line when they are used for a continuity or resistance test, they can only be used in this manner for short periods of time. Therefore, you will need to use a make-and-break technique for this type of testing, rather than continuously holding the meter across a 120-volt power source. FLUKE meters are an exception to this rule. Meters such as the T5-600 or 1000 can sustain continuous contact with rated voltages while on the ohm or continuity setting.

Test purpose. To observe the electrical conductivity of the specific contact under test.

Meter indication.

Contact (continuity) test: A steady tone will sound when the circuit is complete.

Contact (resistance) test: The resistance in ohms will be indicated on the meter display. The reading should be zero. However, a very low reading of several ohms may be permissible.

Test results. Continuity or a specific resistance value can be established for individual contacts under load.

Note: A high resistance value from this test would more than likely indicate the same fault that will be found later when the machine stalls. For the sake of completing this troubleshooting example, the present test results will be ignored.

You have now completed all the work you can do in the panel until the machine malfunctions. During your initial evaluation, you wrote some brief notes in your pocket notebook. Before closing the panel, you review what you have written to be certain that the diagram information you will need later has been properly identified.

Before leaving the machine, you do a general visual inspection of the limit switches you identified on the diagram and of the machine itself. You pay particular attention to the mold protection functions.

As shown in Figure 7-3, this injection molding machine closes the mold at high speed to reduce unproductive cycle time (solenoid E fast mold close, line 17). However, to protect both the mold and the machine, the mold-close function is done at low hydraulic pressure (solenoid D low pressure close, line 15). Near the end of the stroke, an adjustable limit switch LS7 (line 23) slows the mold closing speed (solenoid G slow down line 22). If there are no obstructions, the mold faces will touch at slow speed, at which point limit switch LS11 (lines 15 and 21) changes state, boosting the hydraulic system to high pressure for mold lock-up (solenoid J high pressure line 26). On the other hand, if there *is* an obstruction, the machine will attempt to close at the low pressure setting until timer TR3 (high pressure timer line 27) times out. After timing out, all mold closing hydraulic solenoids will shift to neutral and the machine will idle until an operator opens the mold and clears the obstruction.

You deliberately watch two complete machine cycles to verify that the mold protection functions are working. By watching a hydraulic pressure gauge that is visible to the operator, you verify that the low pressure close (solenoid D) is working properly. By watching the machine movement as the mold faces are meeting, you verify that the slow down (solenoid G)

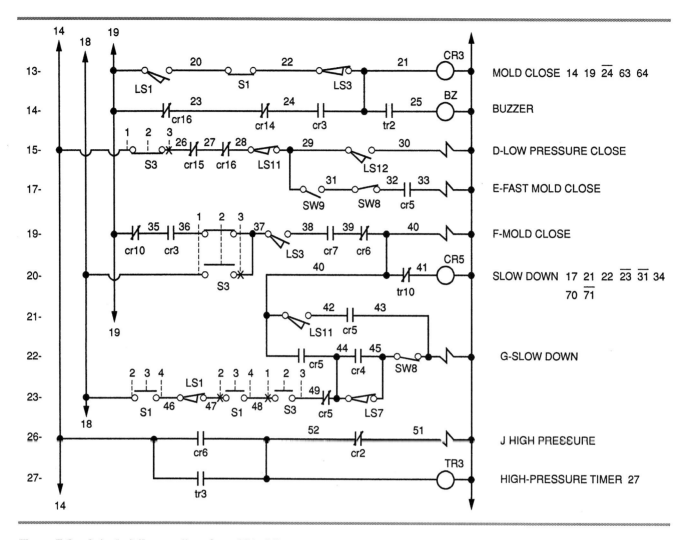

Figure 7-3 Selected diagram lines from 13 to 27.

adjustment has been properly set. *With both of these mold protection functions working, you know it will be safe to allow the machine to continue cycling when you return to the stalled machine for further testing.*

Since everything is in order, there is little more that you can do until you are called. It is now 10:05 A.M. While the machine was running, you took some extra time to study the ladder diagram and use your meter. This will save time later when you are called. You have been by the machine for more than 45 minutes, though actual machine downtime is still at 5 minutes.

You have already spoken with the shift supervisor. She talked with you briefly when you were initially evaluating the problem. You discussed her production schedule and your estimation of machine repair time. She also knows what you want the operator to do if the machine stalls. She will be the one calling you when needed.

Of course, you are going to get another call when the machine stalls. Since we are trying to be as realistic as possible with this example, say that you are paged again at 11:50 A.M. This is realistic because it is 10 minutes before you were going to break for lunch!

When you arrive at the machine you see that the operator followed your directions and left the machine stalled without touching any of the switches or safety doors. You can go immediately to the panel and begin your electrical testing.

ONLINE TROUBLESHOOTING

As you begin the actual on-line troubleshooting, you are careful to follow all of the necessary safety precautions. In addition to being mindful of your own safety when working close to live electrical circuits, you have also verified that the machine could continue operation without danger to either personnel or itself. Remember the caution regarding work on a stalled machine. You must determine that it could safely continue operation before you do any testing. If you electrically complete an intermittently open contact, the machine will continue to sequence. Because the machine failed to eject, the plastic part is still in the mold. However, you have already verified the low-pressure close (solenoid D) and slow down (solenoid G) functions and know that the mold is fully protected should the part fail to fall clear of the mold after ejection.

Reviewing the Known Information

Take a moment now and review the information you already possess concerning the machine malfunction and the subsequent direction you will take in your actual testing. Careful planning and appropriate use of information is an important part of the final time-saving aspect of on-line troubleshooting. Poor troubleshooting techniques are often characterized by a random, haphazard approach to all those mysterious wires in the electrical panel. For the problem that you are now troubleshooting, there is no need to test sections of the circuit that are not related to the ejector function. The failure is in the ejector system—so confine your testing to those circuits.

At the same time, you need to be careful that you do not overlook potential problems because they are not on the ladder diagram line that you are testing. Many times you will find the problem to be in another circuit that is preventing the line you are testing from functioning. A good example would be a failure with the count-up function on lines 37 and 38. A failure on these lines most certainly could result in a problem on line 29 through counter relay contact cr10.

This is what you know about the machine problem when you return to do the final troubleshooting:

1. All main and control power systems are operational because the motor is running and the panel lights and manual controls function. Therefore, you do not need to test fuses or circuit-breakers.

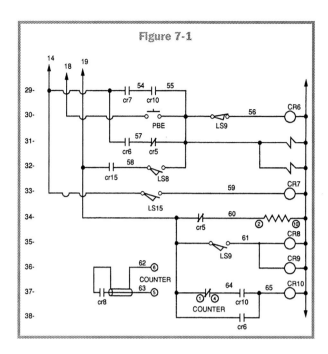
Figure 7-1

2. The specific problem is in the ejector system. However, since the ejector will work on manual, it is safe to assume that the problem is not in the ejector relay CR6, limit switch LS9, or in either of the two hydraulic solenoid valves (L ejector or M ejector pressure).

3. The problem has been intermittent. That suggests that the malfunction will be at some contact point or wire terminal where manual cycling has restored continuity.

4. From your initial testing and study of the diagram, you understand the basic functions of the machine's circuits. Line 32 is a momentary "initiate" circuit while line 31 is the holding circuit. Line 29 maintains the circuit during counting (multiple cycle functions) and the return stroke.

5. As you begin your testing, the machine is stalled, which means that there is an open circuit that should normally be closed. You would expect to find that open-circuit condition on lines 29 to 32.

Continuing with the Actual Testing

Part of the reason you are regarded as a good electrician is your intuitive sense that identifies potential problems during preliminary testing. Your first step after opening the panel is to visually inspect the five relays in the ejector circuit. You find that

92 CHAPTER 7 Practical On-Line Troubleshooting

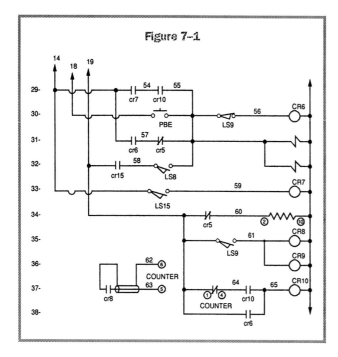

Figure 7-1

cr7 is closed but the rest (cr10, cr6, cr5, and cr15) are open. You also know that there is an initiation circuit on line 32. Since that is a momentary circuit, you know that you cannot test LS8 now. Of course, you could test it by resetting the limit switch, but that would defeat your purpose in having the operator leave the machine untouched after it stalled. You will make those changes later only if you do not find the problem elsewhere, even though the initiation circuit is high on your list of circuit suspects.

Your first hunch, however, is line 31. You think that CR6 is momentarily closing but not holding. The logical test is a continuity (or resistance) reading across cr5's normally closed contact.

Test Point 4

Contact (continuity) test: The meter leads are placed on wire 55 and wire 57.

Non-contact test: Since cr6 is open, this test cannot be used.

Test purpose. To establish the presence or absence of continuity across relay contact cr5.

Meter indication.

Contact (continuity) test: If there were no voltage present, your meter would periodically beep indicating the absence of continuity. However, because the open relay contact will have the full control voltage (120 volts) across the open contacts, your meter will toggle to a voltage display.

Contact (resistance) test: If there were no voltage present, an infinity (0.00 megohm) or very high ohm reading would show on the display. However, your meter will toggle to a voltage display as indicated above.

Test result. The lack of continuity or a low-resistance value indicates that the relay contacts in cr5 on line 31 are defective.

There is a high probability that you have found the reason the machine failed to eject, though you still need to confirm that cr5 is actually the cause of the problem. You could attempt to take a resistance reading across cr5 if you have not already done so. However, you will make your primary verification when you shut the machine down, lock out the power, and remove the inspection cover from the relay contact points.

To finish this example, removing the inspection cover from relay CR5 is exactly what you do. After removing the cover, you find a badly damaged set of contact points, which evidence arcing and heat. You see that the contact spring has collapsed from the excessive heat.

How did you do on your troubleshooting time? It is now 12:04 P.M., which is not good for your lunch break because you still have a relay to either repair or replace. However, from the standpoint of effective troubleshooting, your time is phenomenal. The machine accrued only 19 minutes of downtime for your entire troubleshooting procedure!

Of course, you now need to either replace the contacts if they are available, or replace the entire relay. In either case, you have an inventory of spare parts in the tool room, so the machine will not be down long.

There is one more test you will need to conduct. You still do not know *why* this relay failed. Was the cause simply the relay contact itself, or was there a root cause in another part of the circuit? Since the machine is locked out, your best test is to measure the resistance of the relay (CR6 on line 30) and the solenoids (L-ejector on line 31, and M-ejector pressure on line 32) controlled by *this* relay contact. You will be checking for an excessively low resistance value, which will indicate a shorted or grounded relay or solenoid that is drawing high current. Since the control power is de-energized, you know that cr7 and cr10 on line 29 are open, cr6 on line 31 is open, and cr15 on line 32 is open. Because all of these relays are open and PBE on line 30 is a momentary contact, you know that a resistance reading between wire 55 and the grounded conductor will not introduce any resistance from wires 14, 18, or 19. On the other hand, the two solenoids (L-ejector and M-ejector pressure) and the relay (CR6) are in parallel. You can verify that LS9 on line 30 is closed when the ejector is retracted by doing a quick continuity test between wire 55 and 56.

Since wire 56 comes to a terminal strip, you remove the number 56 wire, which exits the control panel to the machine. This isolates CR6, which allows you to take a resistance reading across the coil itself. The coil reads 48 ohms. After locating a similar relay in the panel, which shows no parallel connections on the diagram, you take a second reading of that coil and discover the two relay coils to be similar in resistance. From this you conclude that there is no apparent problem with the relay coil. Before reconnecting wire 56 to the terminal strip, you measure the resistance between wire 55 and the grounded conductor. The value of the parallel solenoids is 17 ohms. You now find a solenoid on the diagram that has no parallel connections and find its value to be 33 ohms. Since 17 ohms is approximately half of 33 ohms, you can safely assume that both the ejector solenoid (solenoid L) and the ejector pressure solenoid (solenoid M) are serviceable without shorts or grounds. You reconnect wire 56 to the terminal strip and measure the resistance between wire 55 and ground. You now get a reading of 27 ohms, which is the value of the solenoids and relay in series. You have found no indication of a low resistance or shorted coil that was placing a heavy load on relay CR6. You conclude that the machine is safe to put back into operation.

It is time for lunch. You will come back later and do a final test after the machine is back in production.

When you come back to the machine, it is in full production and the operator tells you that everything is running normally. You will now do your final test with an ammeter. You carry a FLUKE T5-1000 in your toolbox. The small fork design of the ammeter allows you to conveniently work in confined spaces. You can safely place the ammeter forks around wires 55 and 56 from the terminal strip. You find that the section of wire 55 feeding the two solenoids is drawing 1.2 amperes and wire 56 is drawing 0.4 ampere when the respective cycles function. Since the control relay is rated at 10 amperes, you conclude that no further testing is necessary. The fault seems to have been confined to the relay contacts themselves with no evidence of external overload.

SPECIAL SETUP TROUBLESHOOTING

On-line troubleshooting lends itself to another variation in the testing procedure. In many cases, you may need to troubleshoot an electrical function that does not cause the machine to stall. A simple and fast way to do this is to reset the machine for a **speed cycle** that allows you to test the faulty function each time it sequences. Of course, the machine could not be producing parts with this setup.

If, for instance, the air system of lines 28 and 29 in Figure 7–4 was not operating, the machine would not stall. However, you could quickly troubleshoot this function if the machine could be taken out of production and the timers changed. Since the air blast is not necessary for the machine's continued cycle but is merely used to blow against the part as an *aid to* ejection, you could set the machine to run without

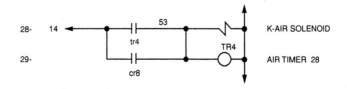

Figure 7–4 Diagram lines 28 and 29.

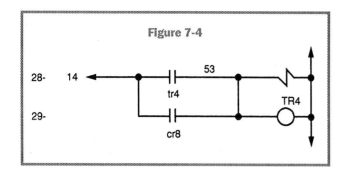

Figure 7-4

this function. Testing would be done on the defective circuit each time it sequences and fails to cycle.

Of course, you could do the testing by running the machine on a normal cycle setting. However, unless there is a need to keep the machine in production in spite of the failure of this system, the simplest way is to do a special troubleshooting setup. Assuming that the machine can be put into operation, you can set the limit switches and timers *so that plastic material will not feed and the various timed intervals are only 1 to 2 seconds in duration.* Thus, when the machine is placed in automatic mode, it will go immediately from sequence to sequence without any waiting periods, because the timed material processing and cooling cycles have been eliminated. The circuit in question can be quickly tested, because it is cycling every 10 seconds or less. This setup procedure can be used on many kinds of equipment where timed intervals or batch weights are a part of the process.

Assuming, then, that you can set the machine up for a speed cycle, you would perform an on-line troubleshooting test on lines 28 and 29 as follows:

1. Record all of the production settings of each timer and limit switch so that you can return the machine to its original setup. You will be working with the machine operator in all changes, both now and when the machine is returned to its original settings.
2. You will set all of the appropriate cycle timers and limit switches to their minimum (or zero) settings so that the machine will cycle at minimum times (and in some cases with zero movement) in each sequence area that you are not testing. You probably will not make changes in areas that will be critical to the test. Before starting the machine, you must carefully check the entire setup for personnel and machine safety during operation.
3. In studying the electrical diagram, you know that if the problem is electrical, it will most likely be a failure of air solenoid K or tr4 on line 28, or the air timer (TR4) or the relay contact (cr8) on line 29.
4. Your first test may be for voltage between wire 14 and the common grounded connector. Since the solenoid and the timer are in parallel, the voltage reading would indicate the completed circuit for both.

Contact (voltage) test: There is a brief audible or visual indication of a voltage during the first portion of the cycle interval.

Non-contact test: There is a brief visual and audible indication of a voltage during the first portion of the cycle interval.

Note: Again you have an initiate and a holding circuit. Relay contact cr8 is a momentary initiating circuit for the timer. Though the same circuit also puts a pulse on the air solenoid, it is too brief to cause sustained solenoid function. However, when the timer is cycled, it maintains tr4 on line 28 and holds the valve open until the timer times out. Thus, you may read two different voltages at the point you are now testing. You may read either a constant voltage if the circuit is fully operational, or a pulse voltage if cr8 is operational but the timer is not cycling. Since there was a brief voltage pulse, you suspect the timer contact is faulty. If you were using a moving-coil VOM for this test, you would see the pointer momentarily swing to the right. Though you would be unable to read a voltage value, the movement of the VOM pointer is very useful for this type of troubleshooting.

5. Your most likely test is now across tr4 between wires 14 and 53. Again, however, because tr4 is parallel with cr8, you will be measuring both functions simultaneously.

Contact (continuity) test: A brief audible tone is heard and the visual display momentarily shifts.

Figure 7-5 Many relays use a mechanical window flag to indicate that they are closed. Some, like these "ice cube" relays, use a neon lamp that greatly facilitates visual troubleshooting of a relay panel. *Courtesy of Siemens Electromechanical Components, Inc.*

Figure 7–5 shows exactly this type of relay. Potter & Brumfield (owned by Siemens Electromechanical Components, Inc.) manufactures relays with a neon light that is visible through the clear plastic case. When the relay is energized, the neon light glows.

There are two advantages to this relay. First, the neon light becomes an effective troubleshooting tool. When the neon light glows, the entire circuit to that relay can be eliminated as the source of the problem. On the other hand, absence of the light—or erratic flickering of the light—confirms an electrical problem on that part of the circuit. Much time can be saved because operating circuits may be eliminated from the items that need to be checked.

This can be carried further than a single relay. Many relay circuits in a typical control panel are dependent on the state of additional relays. Say for example that relay CR5 is dependent on a normally closed contact in CR10, a normally open contact in CR3, a normally open contact in CR7, and a normally closed contact in CR6. If you were troubleshooting this circuit by reading neon lights and CR5 was de-energized, you would look at the neon lights on related relays. You would want to find the neon lights off for CR10 and CR6 because the circuit you are checking uses normally closed contacts in these two relays. On the other hand, you would want to see the neon light glowing in both relays CR3 and CR7. If one of these latter two relay neon lights is not glowing, then you would begin to check on its circuit rather than the circuit of CR5.

Secondly, the plug-in feature of these relays also aids in reducing troubleshooting time. The relays themselves are rarely at fault. Nonetheless, they can be tested by exchanging a known good relay. If the first function now cycles, the problem was a relay fault. Plug-in relays also aid in troubleshooting by allowing functions to be defeated. If you want to test a machine while defeating a specific function, you can do so by removing that function's plug-in relay.

Relays with a built-in neon light initially cost a few dollars more. In the long run, however, they greatly reduce cost by becoming an effective self-diagnostic troubleshooting tool.

EVALUATING THE PROCEDURES

Every troubleshooting problem is unique. There is no standardized test procedure that satisfies every situation. Each of three complete on-line troubleshooting examples given in this book has differed from the others. This includes the test on line 19 given in Chapter 4, and the two examples given in this chapter for the ejector malfunction and the air solenoid test.

The first test on line 19 (mold close) was done by moving from point to point on the circuit and testing for continuity. Since the entire line is a single circuit and is energized for the duration of the sequence, it was possible to do a simple continuity test while the machine was stalled. In other words, the test probes were simply touched to various points along the circuit until the open circuit was found.

On the other hand, the test of the ejection circuit in this chapter involved a number of contact points that were open at the time of testing. Because the lines being tested included an initiation as well as holding circuits, it was not possible to merely test across points on the lines in question looking for a single open circuit. Therefore, even though it was on-line troubleshooting, *it required a combination of* testing of the live circuit and an understanding of the function of the circuit so that the open contacts did not become an unnecessary source of confusion.

Finally, in the brief example of the air solenoid on line 28, you saw a test procedure that involved a machine setup especially for the troubleshooting test. In this last case, more direct on-line troubleshooting with a continuity test was again used.

On-Line Troubleshooting Guidelines

From each of these three examples, there are some general guidelines that can be given for this testing approach:

1. Safe working habits must always be uppermost in the electrician's mind. This is true regardless of the troubleshooting technique used. In on-line troubleshooting, safety precautions must include an awareness of working with live circuits and safety of personnel and equipment either while the machine is running, or in the event that it cycles after being stalled.

2. There will be a great savings in troubleshooting time if an effort is made to understand the circuit and to confine testing to the circuit areas in question. After studying the wiring diagram, you can determine where the problem is most likely to be found. Then conduct your testing so that you systematically move from the most likely to the least likely circuits.

3. You will achieve greater troubleshooting effectiveness when you use your test equipment properly. Take the time to determine the full range of testing capability of the meters you are using, and then take advantage of that potential. Investing in adequate equipment is a necessary first step for the professional troubleshooter who wishes to do truly effective work. Learn to use the best test for the particular need, whether that is a voltage test, a continuity test, a resistance test, or other tests that will be introduced later in the book.

4. Be aware of related circuits. Much confusion can enter into the troubleshooting process if you do not understand the significance of interrelated circuits. If a diagram line is found to be faulty, carefully check all other circuits that are represented on that line. This is the advantage of continuity troubleshooting with an energized circuit. A continuity reading verifies the function of every controlled contact between the test points.

5. Learn to test large sections. A single continuity test from the left side of a circuit line to the input side of the electrical output device or power component, or a voltage test from the grounded neutral to the same point, is all that is needed to verify that a circuit is functioning. If the electrical device or component is not functioning, begin the isolation process by subdividing the circuit into large segments for the preliminary testing. You will save much time by avoiding individual testing of each contact or switch.

6. Generally, most ladder diagram line testing will start with a confirmation of the electrical output device or component on the right side of the diagram, and will then work toward the left of the diagram in the testing sequence.

7. Make judgments prudently as you work. For various reasons, there will be times when you cannot test every device or component in the circuit. Learn how to save time by passing over those areas that would be difficult to test. However, keep track of these shortcuts so that your assumptions do not cause you to overlook electrical faults, forcing you to retest completed work.

Testing with multimeters, whether volt-ohmmeters or digital voltmeters, has not been covered in either this chapter or Chapter 4. Because these meters represent such an important part of the troubleshooting electrician's technique, Chapter 8 deals specifically with testing procedures using these meters.

CHAPTER REVIEW

On-line troubleshooting is most effectively done after thoroughly understanding the equipment's normal function in comparison with its present malfunction. This preliminary information will come from studying the wiring diagram, from a general knowledge background of the equipment and electrical functions, and from specific questioning regarding the particular malfunction. A great deal of time can be saved in the actual troubleshooting process if the objectives of the assigned troubleshooting task are understood beforehand.

Inasmuch as each troubleshooting procedure will have its own unique demands, on-line troubleshooting cannot be reduced to a single formula. Most on-line troubleshooting, however, will fall into one of two formats.

1. The simplest format is one in which the testing is done across a single-line circuit, which would normally be conductive because all of the switches and relay contacts are closed. In this format, on-line troubleshooting consists of verifying the conductivity of the circuit. Any open area constitutes a fault in the circuit.

2. The second format is one in which the testing must allow for open conditions in the circuit during the time of testing, even though the equipment is in operation—or stalled but under power. There may be a number of reasons why the circuit will be open at the time of the test. The circuit may be a momentary or timed circuit, or it may represent an auxiliary or overload function. There may be other reasons as well. In this format, on-line troubleshooting will require that certain parts of the circuit be bypassed or tested using other procedures.

On-line troubleshooting is, by definition, a test on an operating or live circuit. As such, the test procedure will most likely be conducted in one of three machine modes:

1. With the machine running in normal operation.
2. With the machine stalled and unable to continue to the next sequence, but where the controls remain set as though the machine were in normal operation.
3. With the machine speed cycling where the timers, limit switches, and so forth, are set so that the machine can move from sequence to sequence at optimum speed.

For on-line troubleshooting, you will be using either contact testing, which would include voltage tests or continuity and resistance tests with specialty meters, or non-contact testing, which tests a single conductor for the presence of an AC voltage field.

Finally, there are general guidelines that will result in greater effectiveness and speed in your electrical troubleshooting. These include an awareness of safety, understanding the electrical functions of the circuit you are testing, using your test equipment to its greatest efficiency, and conducting your test procedures so that you gain the greatest information from each individual test point.

THINKING THROUGH THE TEXT

1. Give examples of the information you might wish to obtain from a machine operator before starting a troubleshooting procedure.

2. What machine operating conditions would indicate that power supplies (fuses or circuit-breakers) were functioning normally?

3. From the first example given in this chapter, the troubleshooter discovered by pressing push button PBE that, in all likelihood, the limit switch LS9 on line 30 was operational. Why did the electrician come to that conclusion?

4. Aside from electrical hazards to personnel, what is the primary safety concern when doing on-line troubleshooting?

5. From the standpoint of the functions of the machine under test, what is the difference between on-line troubleshooting with normal machine settings and special setup troubleshooting?

6. Testing larger sections of a circuit will usually save troubleshooting time. Explain the meaning of testing larger sections. How will this reduce your troubleshooting time?

CHAPTER 8

Troubleshooting with a Multimeter

OBJECTIVES

After completing this chapter, you should be able to:
- **Understand the necessary precautions for safely using a multimeter in industrial electrical troubleshooting.**
- **Use a multimeter for basic on-line troubleshooting.**
- **Use a multimeter for basic resistance testing within electrical panels, phase-to-ground testing on electrical wiring, and motors or transformers.**
- **Perform specialty tests on capacitors or diodes with a multimeter.**

The use of specialized, solid-state electrical equipment for troubleshooting is an important emphasis in this book. Nonetheless, it would be a disservice to imply that conventional multimeters are not useful instruments for industrial electrical troubleshooting. This is true of the volt-ohmmeter (VOM), the digital multimeter (DMM), and the digital voltmeter (DVM).

GENERAL PRECAUTIONS

When using a multimeter, remember the three precautions that were mentioned in the previous chapters regarding meter safety:

1. Most multimeters have a selector switch that is used to choose the test function performed by the meter. To avoid damage to the meter, it is important that the selector switch be set for the desired test before the leads are touched to the circuit. With the exception of autoranging meters, the function switch must also be set to the appropriate, or a higher, range.

2. Unless otherwise specified by the manufacturer, multimeters can only be used for resistance testing on de-energized circuits. There may be exceptions with a few specialized DMMs. However, all meters with a moving needle (VOMs) must be used on a de-energized circuit when doing resistance testing.

3. Verify your meter on a known energized circuit to ensure that it is operational.

THE BASIS OF MULTIMETER TROUBLESHOOTING

Indication of the presence or absence of a voltage is the multimeter's most useful function in industrial troubleshooting. A voltage can be used to determine a live or functioning circuit, or it can be used to indicate an open condition along a conductor. Secondarily, the multimeter can be used for either continuity tests or actual measurement of resistance values when the circuit is de-energized.

Troubleshooting with the Multimeter

In Chapter 7, you were troubleshooting the ejector system. Go back and redo a part of that procedure with a multimeter. As you did before, start by collecting preliminary information. From the preliminary testing, you have already determined that the problem is with the ejector system on lines 29 through 32 shown in Figure 8–1. Now, with the machine running, you will be testing to identify the circuit's functions.

Voltage Testing

Your first test procedure will be done on the meter's voltage ranges. You are attempting to find a point in the circuit at which the conductor is open.

Test Point 1

The meter leads are placed on wire 55 and the common grounded conductor.

Selector switch setting. 200 VAC.

Test purpose. To determine the presence of a control voltage.

Meter indication.
VOM indication. The needle swings to the right as the circuit is energized.
DMM indication. The display counts up, indicating full line voltage as the circuit is energized.

Test results. The circuit is verified as operational. A control voltage is indicated for the full duration of the ejection cycle.

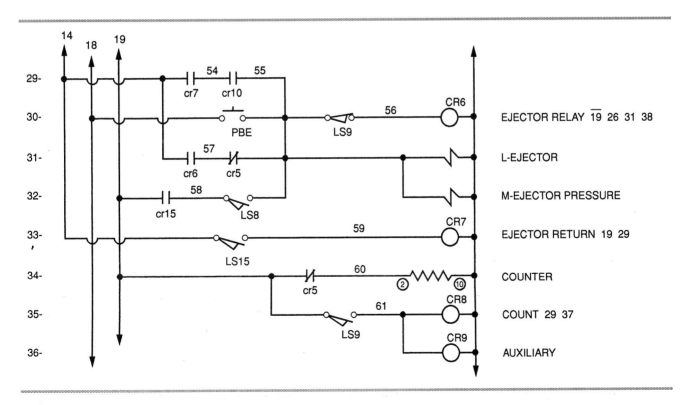

Figure 8-1 Diagram lines 29 through 36.

Test Point 2

The meter leads are placed on wires 55 and 14.

Selector switch setting. 200 VAC.

Test purpose. This test will establish the sequence of the control circuit feeding CR6 and solenoids L and M. A single meter reading will be indicated irrespective of which of the three lines (29, 31, or 32) is conducting.

Meter indication.
VOM indication. The needle will indicate 120 volts when the combined circuits of lines 29, 31, and 32 are open and LS9 is closed. When any one of the three circuits is energized, the needle will swing toward zero volt.
DMM indication. The display will indicate 120 volts when the combined circuits of lines 29, 31, and 32 are open and LS9 is closed. When any one of the three circuits is energized, the display will count down toward zero.

Test results. Under normal cycle conditions, the voltage drops to zero during the entire ejector cycle time. The test indicates that there is at least one closed circuit on line 29, 31, or 32 during the entire ejector cycle time.

As you saw in Chapter 5, when you are using a multimeter to check a live circuit for conductivity, you must read the information correctly. As in the example above, if you are reading across an open contact on an otherwise complete energized circuit, the meter will show full line voltage. When the circuit is closed, the voltage difference between the two points should drop to zero. The presence of a voltage less than the control voltage may be an indication of a high-resistance contact or wire connection. If you were reading across cr6 on line 31 while it was under load and your meter showed from 2 to 5 volts, it would indicate that the contact was failing. On the other hand, a voltage loss across a closed contact in good condition might be less than a meter set for the 120-volt control circuit would readily indicate. Your multimeter will therefore indicate one of three conditions when you place the leads across a contact: (1) A full control voltage reading indicating that the contact is open. (2) A zero voltage indicating that either the contact is closed or that the part of the circuit that you are testing is de-energized. (3) A voltage substantially lower than the control voltage indicating that the contact has high resistance and is failing or that you are **backfeeding** through other circuits. The difficulty, however, is in distinguishing between a closed contact and a de-energized circuit. For example, if you are reading across wires 14 and 55, you will get a zero reading irrespective of the conductivity of the wires if LS9 is open. Thus, if you have a zero reading, it is necessary to verify the control voltage by testing between wire 14 and ground. A zero voltage between wires 14 and 55 and a 120–volt reading between wire 55 and ground indicate that the circuit between wires 14 and 55 is conductive.

You will frequently encounter backfeeding when testing on live open circuits with a DMM. A circuit that is open between its primary power source and neutral connection can be fed through other connections that are common to the power source. However, because of the resistance of the common circuits, the voltage reading may be from a few volts to 80 volts or more. Typically, however, that voltage would not sustain sufficient current to do any useful work. Nonetheless, your meter will display the voltage value and you must account for it.

A potential backfeed condition from Figure 8–1 can be demonstrated with a machine setting that would not normally occur. However, if the machine was in a manual operation mode, wire 19 would be de-energized and wire 18 would be energized. If PBE on line 30 was depressed and held, wire 55 would become energized. Assuming that LS8 on line 32 was closed, were you to read across an open cr15 on line 32 by placing your meter probes on wires 19 and 58, you would obtain a reading of some partial control voltage value. Wire 58 on line 32 would be carrying 120 volts relative to ground. However, since wire 19 is open, it becomes a high-resistance conductor back to ground. Figure 8–1 shows at least one connection to ground on wire 19 through the normally closed relay cr5 and the counter on line 34. Since the counter probably has a high-resistance solid-state circuit, the backfeed voltage will be small. On the other hand, if relay LS9 on line 35 was also closed, the coils of CR8 (line 35) and CR9 (line 36) would also become conductors between wire 19

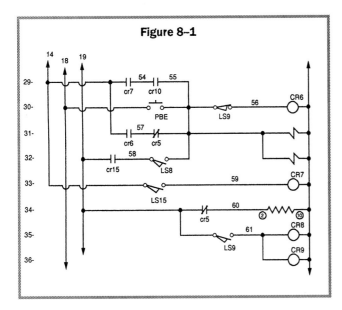

Figure 8–1

and ground and the voltage reading would approach 120 volts because of the relatively low resistance of the relay coils.

In the troubleshooting example on lines 29 through 32 from Chapter 7, you did some additional testing while the machine was running. Similar preliminary testing could be done with a multimeter as well, though you will not go through each of those steps here.

On-Line Testing of a Stalled Machine

After the machine stalls, you will again start your on-line testing procedure. As suggested in the previous discussion, you suspect that cr5 on line 31 is failing. That is, again, how you will conduct your multimeter testing.

Test Point 3
The meter leads are placed on wires 55 and 57.

Selector switch setting. 200 VAC.

Test purpose. To verify that the normally closed relay contact cr5 on line 31 is conductive.

Meter indication.
VOM indication. 120 volts AC.
DMM indication. 120 volts AC.

Test results. Relay contact cr5 has failed.

The presence of a voltage indicates that there is an open conductor at cr5. The meter is reading a voltage on two conductor ends (wires 55 and 57). If cr5 is open (but all other contacts on the line are closed), then all contacts including the electrical output devices or components (the wire-wound coils in CR6, solenoid L, and solenoid M) act as conductors. Rather than having a voltage drop across the electrical output devices or components as it should, the full voltage drop is across the open part of the circuit. For this reason, the open part of a circuit will always read full line voltage if the rest of the circuit is complete.

Off-Line Resistance Testing

At this point, you could verify your findings with resistance tests. You would need to shut the machine down for your test since a resistance test with a standard multimeter must be done on a de-energized circuit. However, since cr5 is normally closed, this will not alter the relay contact positions. You could not do a resistance test on a normally open contact with a standard multimeter, since de-energizing the circuit for the test would open the relay. You could, of course, de-energize the circuit and manually close the relay, but that could alter the performance of an intermittent problem.

Your first step is to de-energize the machine. This would most likely be done by opening and locking out the circuit-breaker. A wise precaution before using your multimeter as a resistance meter would be to measure for a voltage between ground and the main control wire 14. If there is no indication of voltage, you can proceed with your testing by setting your meter on a low-resistance range.

Test Point 4
The meter leads are placed on wires 55 and 57.

Selector switch setting. The meter is set on either the 2– or 200–ohm resistance range (or the X1 range on a VOM).

Test purpose. To test the resistance (measured in ohms) across the normally closed contacts cr5.

Meter indication.
VOM indication. The needle indicates infinity (∞) or a very high resistance value.

DMM indication. The display indicates infinity (0.00 MΩ) or a very high resistance value.

*Note: Correctly reading a DMM value will take a little practice because there is less standardization of readout symbols. Each manufacturer uses what best suits its purposes. Second, with autoranging meters, such values as infinity (∞) will be designated with existing **alphanumeric** characters and decimals. Thus, infinity may be indicated as a function of **megohms** (0.00 MΩ) or even as a combination with the half-digit function displayed as 1.00 megohm. You will need to familiarize yourself with your own meter's symbols.*

Test results. The normally closed relay contact cr5 is defective.

As you can see from the example you just completed, the multimeter (VOM or DMM) can be used effectively for industrial electrical troubleshooting. Even though it cannot be used for continuity testing in its resistance ranges on live circuits, it can perform many of the tests described as on-line troubleshooting.

SPECIALIZED MULTIMETER TESTING

In the following sections, you will see some testing procedures that go beyond the circuit of the ladder diagram. These tests will be done with a multimeter.

In many cases, absolute electrical values are not the objective of the multimeter testing done in industrial electrical work. If, for instance, you are measuring the resistance of each of the three windings in a three-phase motor, it is probably insignificant if the value is 3 or 7 ohms. What is significant is that there is a reading indicating a complete circuit with no reading to ground. It is also significant that the winding resistances are balanced; each of the three-phase coils must have equal resistance. The same is true of most voltage readings. With exceptions such as reading absolute voltage values to compare phase balance, most voltage readings done with a multimeter in industrial electrical work are done for the purpose of determining the presence or absence of an approximate voltage. With that in mind, there are a number of specific troubleshooting techniques that use the multimeter.

Multimeter Resistance Testing

The following tests can be made with a multimeter on the resistance settings. Consequently, all of these tests are done on de-energized circuits. It will not be repeated for each section, but it goes without saying that the first test would verify that the circuit is de-energized, both as a safety precaution for personnel and as a prerequisite for resistance measurement with a multimeter. Verification of a de-energized circuit is done with a voltage range higher than the known voltage of the system under test.

Most resistance testing for industrial electrical work is not concerned with specific resistance values, but is rather looking for the presence or absence of a resistance that would be significant for that particular test. In other words, a test between a conductor and ground should show no appreciable value on the meter's lower resistance scales. Infinity (no resistance) would be an acceptable test result, whereas any significant resistance value would indicate a short-circuit. A test of a magnetic starter coil would not be concerned with the actual ohm value as long as it showed continuity. It could be as high as 400 ohms or as low as 10 ohms, depending on the voltage rating and size of the coil. For this reason, most resistance testing in industrial work will be on the low-resistance ranges. The exception to an almost exclusive use of a low-resistance range is insulation testing, though that is more accurately done with **megohmmeters**. This will be discussed in Chapter 9.

Phase-to-Phase and Phase-to-Ground Testing

The multimeter is often used for short-circuit testing between two phases (hot legs) or a phase leg and ground. For example, if you were testing a remotely located three-phase motor that was tripping its circuit-breaker, it is possible that you would want to test for shorts between conductors (**phase-to-phase**) or for shorts to ground (**phase-to-ground**). For the phase-to-phase test you would need to isolate each of the lines. Since there is a circuit-breaker at the main panel area, and a disconnect within sight of the motor, you can open and *lock out* both the circuit-breaker and the disconnect, and the lines between the circuit-breaker and motor disconnect will be isolated. To test for a phase-to-phase short in this de-energized, isolated section of line, you will use your multimeter to test between phases A and B, between phases B and C, and between phases A and C. Using the low resistance range on the meter, you would be looking for any resistance value or needle movement as an indication of a short between the conductors.

Phase-to-ground is an easier test because you do not need to isolate the leads. Obviously, however, you do need to disconnect the power. For the same remote location motor, you would test between phase A and ground, phase B and ground, and phase C and ground. If there was no display indication in any of these three tests, you would conclude that the motor, or its **feeder** circuits, was not shorted to ground.

If you understand how the multimeter was used to test the feeder lines for phase-to-phase or phase-to-ground shorts, you will be able to apply that same testing procedure to many other pieces of electrical equipment. Motors and transformers are often tested for shorts in the same way. The phase-to-ground is again the simplest test. You would test between any line lead and the motor frame. Any significant meter reading would indicate a short. To test for phase-to-phase shorts, you will need to isolate each set of windings and test the leads between any two sets of coils. Again, the absence of meter movement is an indication that there are no short-circuits. In this case, you would want to use higher resistance ranges on your meter. However, with a standard multimeter, the lower ranges will indicate a reading with a shorted motor, whereas the higher ranges may not indicate a fault even if the motor insulation is breaking down just before complete failure.

Transformer testing is done in much the same way as the motor example above. With an **isolation transformer** in which the primary and secondary are separate windings, there should be no resistance reading between the high (primary) and low (secondary) sides of the transformer. Similarly, there should be no reading between any of the leads and the transformer case. The exception is if the transformer is used as a control transformer with a grounded conductor and is still in the circuit. Then one leg of the secondary circuit is grounded. A system ground may also be common to one side of the primary, which would also give a reading. However, in both cases, if the leads are disconnected from the transformer, there should be no resistance between ground and any lead.

Continuity Testing

Resistance is a function of a completed circuit. For this reason, the multimeter's resistance ranges are ideal for verifying electrical circuits. The simplest example would be a magnetic motor starter coil. In the case of a nonfunctioning magnetic starter, you would need to verify the condition of the coil. Taking a reading across the coil on a low-resistance range would indicate infinity (for an open coil) or low resistance (for a good coil). The precise resistance value is of little significance since you are merely testing to see if the coil is open or good.

Many similar tests will use the resistance settings on the multimeter. All of the output devices or power components on the right side of the ladder diagram may be tested in this way. A relay coil, a hydraulic solenoid valve, a timer motor, or an indicator light would all indicate resistance if the circuit was complete, and infinity if the circuit was open. Ground short tests can also be made by testing between a power lead and ground. Remember, however, that unless you disconnect the grounded side of the electrical component, you must allow for the internal resistance of the device or component itself. Figure 8–2 demonstrates the use of resistance values for determining the condition of a coil. We will assume that all coils in this figure have similar resistance values, and we will read the resistance between their respective terminal wire and ground. Coils A, B, and C are good coils. A resistance reading between coil A and ground will indicate the normal resistance value for the coil. Coil B/C is in parallel. The resistance value of coil B/C is approximately one-half that of coil A. Coil D has an open conductor between the coil and the common grounded conductor. The resistance value of coil D is infinite. Coil E is internally shorted between coils. The resistance value of coil E will be *less* than that of a normal coil. Coil F is internally open from either physical damage or overloading. The resistance value of coil F to ground is infinity.

Similarly, the ohmmeter can be used to verify switches. The meter leads can be placed across

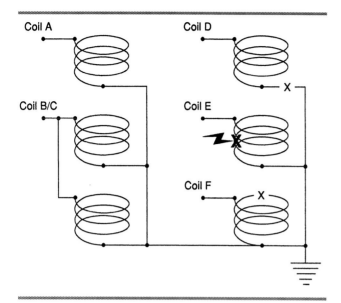

Figure 8-2 A resistance reading between coil A and ground will indicate the normal resistance value for the coil. The resistance value of coil B/C is approximately one-half that of coil A. Coil D has an open conductor between the coil and the common grounded conductor. The resistance value of coil D is infinite. Coil E is internally shorted between coils. The resistance value of coil E will be *less* than that of a normal coil. Coil F is internally open. The resistance value of coil F is infinite.

switch terminals to verify that the switch is functioning. Again, you may need to allow for higher resistance values if the circuit can backfeed through other parts of the circuit.

When you are testing resistance with an ohmmeter, you must be certain that the circuit you are testing is isolated. The output device or component being tested cannot be in parallel with any other closed circuit. Your meter will measure any resistance that is in the circuit. For example, if two relays are connected in parallel as shown in Figure 8–2, you will obtain a reading whether both or only one coil is good. Consequently, if you are attempting to test a coil that is normally in parallel with another, you will need to disconnect one of the leads to isolate the coil before doing the test.

There is a shortcut, however, that can sometimes save you the time of isolating output devices or components on a parallel circuit. Measure a similar component in another part of the circuit—such as a relay coil—for its resistance value. Since the value of parallel resistance is less than the value of a single resistor, a resistance value of less than that of a single similar component indicates that both components in the parallel circuit are functional. If the two devices or components have the same resistance value, the combined value will be half. For example, if you measure another similar relay coil and find its value to be 100 ohms, the two identical parallel coils would read 50 ohms if both were functional. If you obtained a value of 100 ohms, you would know that one of the coils was open. If you obtained a value considerably less than 50 ohms, you would know that one of the coils was shorted to ground on the hot side.

Lead or Conductor Identification

A multimeter can be used for lead or conductor identification. For example, if you have a **step-down transformer** with a 240–volt primary and a 120–volt secondary, which has four leads but no marking, you could identify the lead sets, but not winding function, with your multimeter. You would test between one lead and any of the remaining three. When the meter showed a resistance value, you would have identified the two leads to one coil. The remaining two leads would then test as the second coil.

Frequently, when testing in an electrical panel, numbered markers will be missing from wire ends. If, for instance, you know that wire 54 goes between relays CR10 and CR7, but the markers are missing from three of the wires leading to CR7, identification is simple. You will set one meter lead on wire 54 on CR10 and subsequently touch the second lead to each of the unmarked wires on CR7. When the meter indicates a reading, the unmarked wire 54 has been located. Of course, you must verify that relay contacts or switches common to the wires you are testing are all open so that you are not reading through other circuits.

Capacitor Testing

Though there are specialized capacitor testers, the multimeter can do a satisfactory job identifying low-value shorted capacitors. Before testing any capacitor, however, short the leads with a resistor and allow the capacitor to drain. Electrolytic capacitors can hold high charges that could damage a meter. The first test will determine if the capacitor is shorted. A good capacitor should show an infinity meter reading when the test leads are touched to the capacitor terminals, whereas a shorted capacitor will show a low-value resistance.

Another interesting test for electrolytic capacitors gives an indication of their capacitive function. You must use a VOM for this test. Start the test with the meter on the X1 range—higher resistance ranges give greater needle deflection, which may exceed the scale range of the meter. Place the test leads on the capacitor terminals. The needle will rise slightly and then fall because you are actually charging the capacitor with the meter's battery. Now, reverse the leads and watch the needle. The needle should swing sharply to the right and then slowly fall as the capacitor discharges. If the needle does not swing, the capacitor is leaking internally. If you do not obtain satisfactory readings, you may need to use higher resistance ranges on the meter.

Note: These tests can be done on low-value capacitors. High-value capacitors and line capacitors should not be handled without proper training.

Diode Testing

A multimeter can be used for a number of solid-state tests. The electrician in industrial electrical work frequently checks **diodes**. A properly functioning diode will block the current in one direction, while allowing it to pass in the other. Thus, for the first test, the meter leads are placed across the two diode terminals. The meter should indicate either a very high or a very low resistance value. By reversing the leads, the opposite (either low or high) reading should be indicated. A good diode will indicate a low resistance in one direction and a high resistance in the other. A faulty diode will most often indicate a low resistance in both directions, though a high resistance in both directions is also possible. Most DMMs have a diode-testing function. Usual indication of a good diode is obtained with an audio signal in one direction and no audio signal when the leads are reversed. In some instances, one end of the diode circuit must be opened for accurate testing.

The diode test uses the blocking effect of the diode as the means for devising test results. A multimeter uses a direct current flowing from the negative probe to the positive probe in all resistance tests. The resistance being measured is the conductor between the two probes. When a diode is tested, the current will flow through the diode in one direction, but will be blocked when the test leads are reversed. If the diode is damaged so that it is internally open, a current will flow in neither direction. If the diode is internally shorted, a current will flow in both directions.

There is a potential hazard when using multimeters with solid-state equipment. Many solid-state circuits can be damaged by even the low current output of a moving-coil VOM. A VOM should never be used on circuit boards with integrated circuit chips or other low-power solid-state devices. Generally, a DMM that has a much lower current output can be used for this type of testing, though it is your responsibility to determine circuit safety before using any multimeter on solid-state devices.

Industrial electricians seldom use a DMM on resistance ranges in hundreds of thousands to millions of an ohm. Nonetheless, there is a worthwhile practice to establish when measuring high-value resistances. Holding one test lead in each hand with sweaty fingers can produce a reading from 8 megohms (eight million ohms) to as little as 3 megohms. If you hold any high resistance item in your hands and touch the leads with your fingers, you are adding your own skin resistance in parallel to the device you are attempting to measure. If you are checking a high-current welding diode, adding a few million parallel ohms will have no consequence on the reading. Nonetheless, as industrial electricians we must occasionally hold small electronic devices in our hands while doing resistance measurements. Get in the habit of holding the device and the test lead between two fingers with one hand and holding only the insulated portion of the lead with the other. Use the insulated lead to touch the second terminal with no support from your hand.

EVOLUTION OF THE DMM

Manufacturers are making constant improvements in electrical test instruments. When the first edition of this book was written, there were very few test instruments capable of continuity testing on live circuits. Even then, those functions had been included only because their manufacturers wanted to protect the meters from overvoltage in a resistance setting. Presently there are a substantial number of meters capable of true on-line troubleshooting testing.

The FLUKE 12B multimeter, which was shown in Figure 5–6, is an excellent example of present-day meter development. The 12B was introduced as an inexpensive meter, yet it performs exceptionally well as an on-line troubleshooting meter. The 12B is shown in use in Figure 8–3.

You can use the FLUKE 12B or the similar FLUKE 16 meter in the resistance function on any voltage up to 600 volts alternating or direct current. Whenever the meter encounters more than 4.5 volts alternating or direct current while in the resistance function, it automatically toggles to a numeric voltage display. Consequently, you can use the resistance range for on-line testing and the display will read precise numeric values for either volts or resistance without changing function. Resistance values will read to 0.0 ohm resolution and voltage values will read to 0.000 volt resolution.

Imagine the information you will have available with this meter when you are troubleshooting across a set of energized relay contact points in an electrical panel! If the points are in good condition, you should see a value of less than 0.2 ohm. If you see values greater than 0.5 ohm, you probably have contacts that are failing even though the machine appears to be operating normally. Remember that Ohm's law also applies to relay contact points. As the resistance increases, the voltage across the points will increase if the load remains constant. At some time, the voltage drop will be high enough that the FLUKE 12B meter will automatically switch to a voltage display. In a rapidly failing relay you may get a fluctuating voltage value from tenths of a volt to 1 or 2 volts.

There are no absolute resistance or voltage drop values to indicate contact suitability. You will need to check similar circuits to determine what values other relays with the same load show. However, each test point you now make with the 12B meter will give you a numeric value on which to make your diagnosis. In the example of Chapter 7, you waited until the machine stalled before you could do your final on-line troubleshooting. In all likelihood, with this meter you could have found a high resistance (or voltage) value when you tested across the weak relay contacts. The resistance value (or voltage value) would have drifted considerably during the time the relay was closed. For example, you may have seen the resistance fluctuate between 0.2 and

Figure 8-3 The FLUKE 12B in use. This meter will give numeric values for either voltage or resistance without changing the function setting. *Courtesy of FLUKE Corporation.*

3.0 ohms. With this information, you could have saved the entire second step of troubleshooting a stalled machine.

The FLUKE 12B has a second useful function for troubleshooting intermittent control problems. FLUKE calls it Continuity Capture™. The meter can be set to record 250 **milliseconds** or greater make-or-break events in continuity. Figure 8–4 shows how the transition appears on the screen. For example, limit switches can become worn or out of adjustment to the point where they randomly open. Using the Continuity Capture display, you could record the pattern shown in the left side of Figure 8–4 if a limit switch was momentarily failing during production. This meter function has great potential in on-line troubleshooting. It essentially gives a specialized **trend plot** oscilloscope function to an inexpensive DMM. You will thoroughly evaluate the trend plot in Chapter 10.

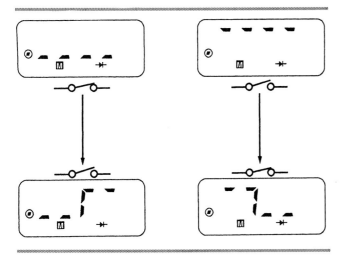

Figure 8-4 The FLUKE 12B Continuity Capture display, which shows a contact change from either open to closed or closed to open. *Courtesy of FLUKE Corporation.*

Figure 8-5 The Tektronix TX3 DMM shown connected to a laptop computer. This figure demonstrates a real-time test with the DMM connected to an energized circuit. More complete test data is numerically displayed on the sidebar of the computer screen than can be viewed on the DMM itself. The DMM function can be controlled remotely from the computer keyboard. Note the RS232 cable connected to the top of the meter. This is an optical interface that isolates the computer from transients that may travel from a live circuit to the meter. *Courtesy of Tektronix, Inc.*

In addition to both toggling between absolute resistance and voltage values and the Continuity Capture function, this DMM has full voltage reading capability to 600 volts. It tests both diodes and capacitors. In addition, it can be used to record minimum and maximum voltage or resistance values. It also has an elapsed time function that can record a minimum or maximum event to 99:59 hours.

As noted in Chapter 5, however, this meter has higher impedance in the ohm range and is capable of drawing enough current to close an ice cube relay. You must verify that this will not create any hazard during testing before using the ohm range on an energized circuit, because the meter could cycle small relays by completing a circuit.

COMPUTER-ENHANCED MULTIMETERS

A DMM or other meter can become a powerful test instrument when it is linked with a computer. Many manufacturers are producing their top-of-the-line test equipment with both hardware and software so that the meter can be interfaced with a computer. Figure 8-5 shows Tektronix's TX3 DMM (or TX1) linked with a laptop computer for field testing and recording. The following discussion is primarily concerned with DMMs. However, this material can be equally applied to any hand-held electrical meter linked to a computer. In Chapter 10 we will consider computer-linked oscilloscopes.

The Tektronix TX3 is a true RMS meter. It has a full range of functions, including DC/AC voltage (500.0 millivolts to 1,000 volts), DC/AC current (500.0 milliamperes to 10.00 amperes), resistance (50.00 ohms to 50.00 megohms), capacitance (5.000 nF to 50.00 mF), **frequency** (0.5 hertz to 1 megahertz), and temperature (50°C to +980°C). With a current probe the TX3 can measure and record up to 500 amperes AC. It has a 1,000–volt CAT III input voltage safety rating. The accompanying WSTRM PC interface package is Windows® 95/NT® compatible.

Advantages of Computer Linking

The portability of laptop computers coupled with hand-held meters produces powerful test and recording instruments for industrial plant maintenance. In the years to come, this will become a commonplace application in advanced industrial settings. The use of the computer increases the meter's capability in the following areas:

1. *Enhanced waveform and numeric displays.* The computer is used to generate larger and more readable graphics. In addition, the computer is often used to give numeric values in sidebar information, which either is not available on the meter screen, or is more difficult to read with precision.

2. *Permanent storage of test data.* All manufacturers of computer-enhanced meters use the computer to increase the storage space for captured information. The TX1 can store ten entries in its nonvolatile memory and the TX3 can store 30. However, with the use of a linked computer, the storage capacity is limited only by the computer memory. With the extended storage capacity of the TX3, much information can be carried in the DMM itself until it is convenient to download it into the computer. Some manufacturers' meters have a memory capacity of as few as a single event to three events.

3. *Capability to graph and print test data.* The information on the meter's screen or the enhanced graph from computer memory can be printed for permanent records or analysis. Even though FLUKE has a hardware package that allows the screen data to be printed directly from certain meters, the results are more complete when the same information is printed after enhancement through an interfaced computer.

4. *Remote control of the meter.* Tektronix's TX1 and TX3 and FLUKE's 89 series 4 DMMs allow function control of the meter from the computer itself. Undoubtedly, in time more meters will allow function change from the laptop keyboard.

5. *Remote control of the process.* At this point, we know of no computer-enhanced meters that allow control of the process from the linked computer. However, this is a logical next step that manufacturers could provide. If the meter can monitor functions such as voltage or amperage peaks or sags, it is merely a programming function to allow the computer to output a signal when the value is above or below operator-controllable set points. Of course, output hardware would also need to be added to the computer itself.

Not all meter software packages have the same features. In almost all cases, meter manufacturers use the computer for the first three categories: to enhance waveform or information display, as a means of permanent storage, and as a tool to print the recorded information. In the case of the TX3, these features are well developed and powerful. Fluke's software packages are equally well developed for their computer interfaced test equipment. On the other hand, some manufacturers include similar features with far less usefulness. *You must carefully evaluate each software package to be certain that it does what you expect.* Remote control of the meter itself is currently limited to very few meters. Finally, using the meter itself to remotely control the process is still in the future.

Computer Linking with Optical Isolation

A hand-held meter will probably use one of two interface links with a computer. First, the communication line may be a hard-wired output that directly connects the logic circuit of the meter with the communication port of the computer. Alternately, the communication link may pass through an **optical isolation interface** so that there is no direct electrical connection between the circuit board of the meter and the computer. Both of these communication links may be listed as **RS232** inasmuch as this standard specifies the maximum *voltage* rather than the *isolation* of the communication link.

With a little thought, the hazard of using a direct electrical connection between a meter on an energized circuit and a valuable laptop computer should be apparent. If any transient crosses between the input circuit and the communication output of the meter, that same transient will be carried to the communication input of the computer. At the very least, the computer will be damaged. On the other hand, if there is an optical interface in the communication link between the meter and the computer, the computer will not be damaged even if the meter is entirely destroyed by a transient.

High-quality meters such as Tektronix's TX series and FLUKE's computer-interfaced meters use an optically isolated RS232 communication cable. Look again at Figure 8–5 and you will see the large "plug" on the cord attached to the meter. If you remove the cord and look at this "plug," you will see two optical lenses. There is no metallic (electrical) connection between the meter and the computer. A transient cannot cross the optical barrier. On the other hand, meters are available from other manufacturers that use phone jack connections between the meter and the communication cable to the computer. If this represents a direct electrical connection between the meter and the computer, the computer is in jeopardy from transients.

It is possible that a well-protected meter could use an optical isolation interface built into the meter itself. All of Tektronix's hand-held TekScopes® use their IsolatedChannel™ architecture, which optically isolates the energized channel inputs from the meter's logic circuit. In this case, a phone jack connection is used between the meter and the computer, though the output to the computer interface is optically isolated from the live circuit. Nonetheless, the rule of thumb is that a meter can safely be used on energized circuits while connected to a computer only if you can visually verify the optically isolated connection cord. If you cannot visually verify optical isolation, then you must check with the manufacturer to confirm that your meter is internally optically isolated. If the meter is *not* optically isolated, it should only be connected to the computer when downloading or manipulating stored information. *You should not connect a hand-held (battery powered) meter simultaneously to a live circuit and a computer if the meter does not have either internal or interface cord optical isolation.*

Practical Applications of Computer-Enhanced DMMs

For these examples, assume that you are the maintenance chief of a seafood processing plant. When the catch is unloaded, it is mandatory that all conveying, processing, and refrigeration equipment operate at full capacity. Downtime of any equipment causes extreme hardship for the line crews and can potentially cause expensive product loss. Even though this plant uses a relatively small number of process machines, your company has developed a thorough preventative maintenance program. Because of the high humidity and salt air, you test major motor contactors monthly and critical refrigeration current weekly. To avoid transcription errors, you download the recorded data directly from your meter to the computer in your office.

Test Example #1

The processing line has four fish-skinning stations. Each station uses a 5–horsepower, 240–volt, three-phase refrigeration unit to surface-freeze a rotating stainless steel drum. The filleted fish is stripped as the skin adheres to the freezing surface of the drum. According to both the nameplate and your own observation, each hermetically sealed compressor motor draws 15.2 amperes. Each week after the processing lines have been operating for at least one hour and the incoming line voltage is stable, you record the current for each refrigeration unit from the central motor control center. You use a Tektronix TX3 with a 100–ampere current probe. Because the TX3 is capable of resolving 15 amperes to two decimal places (15.00), you have noticed that unit #3's readings are progressively showing higher amperages. They have changed from 15.2 amperes to over 15.7 amperes in a period of 25 weeks. Even though the 1.15 service factor allows the motor to run at 17.48 amperes, the slope of the plot warns you that it is time to do further testing. You will want to verify both line voltages and establish a trend plot with a **megger**. See Figure 8–6.

Test Example #2

The central motor control center supplies the four refrigeration units, two freshwater and one scavenger

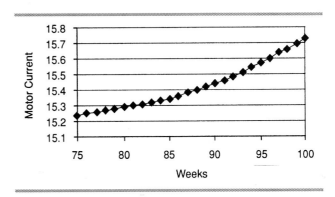

Figure 8-6 A weekly current plot taken on a failing 5-horsepower, 240-volt, three-phase motor. The readings were taken with a Tektronix TX3 digital multimeter using a current probe and then downloaded to a computer for permanent storage. *Courtesy of Tektronix, Inc.*

pump, a large walk-in freezer, the conveyor systems, and multiple processing equipment controllers. Both experience and the priority of each type of equipment have dictated a schedule for verifying the condition of certain motor contactors. You have determined that a monthly log of contactor condition should be kept for each refrigeration unit because the rotating refrigerant unions on the drums result in high-maintenance problems for the entire unit. Because of their importance to production, the main water pumps and their electrical contactors must also be frequently monitored. Conveyor motors and small process equipment are not supervised as closely. Your TX3 has several functions that allow quick and accurate verification of the motor contactors.

On-Line Testing. Your test of preference is an on-line verification of the contacts under load. Because your TX3 has the ability to measure millivolts with resolution to 000.0, you are able to record precise voltage loss over each contactor. After the meter is set in the voltage range, you can zero the display to 000.0 millivolt. During the actual test across the closed and energized contacts, the display value will indicate the voltage loss across the contacts in millivolts. You can then press the HOLD button, enter the memory function, and store the reading in the next available memory location. With your notebook in hand, you identify each memory location by number with its corresponding contactor as you make the measurements. Later, when you download the information into your computer, you will be able to identify each reading with its appropriate contactor. Your TX3 can store thirty measurements before you must save it to your computer.

Off-Line Testing. You could also test the contactors with the equipment off-line. In this case you would de-energize and secure each line disconnect in its open position. However, you would leave the control circuit energized so that you could close each contactor with its manual function switch. The de-energized contactors could be closed to take a resistance reading across the contacts. Your TX3 has a 50–ohm resistance function with 00.00 resolution. This precise resistance capability will show very small changes in contact wear. You could proceed with your resistance testing of each contact and record it for eventual downloading into your permanent computer memory.

The value in testing and recording either millivolt or ohm values across contact points is not in the precise values themselves. There is no pass-or-fail value at which you must change a contact. Rather, you are looking for a trend where values rapidly begin to change. The meaning of this change is discussed more completely in Chapter 10.

Test Example #3

You are replacing an older processing machine with a new unit you have just installed. The new unit connects a 20–horsepower motor to a gearbox with a 90–volt direct current clutch. The unit operates satisfactorily with the exception of the clutch. You have noticed that the clutch carries full-load when the control panel doors are open. However, with the control panel closed, the clutch will occasionally slip after an hour or more of operation. The problem, however, is not in a lack of panel ventilation, because it is in a wash-down area and the system was designed with a nonvented panel.

You suspect that the problem is in the 90–volt direct current **power supply** for the clutch. However, since you cannot do the testing with the panel doors open, you decide to enclose your TX3 in the control panel and read the values from a laptop computer. The gasket on the panel door will allow you to bring the RS232 cable out without damage while the door is lightly closed.

However, you need to test values at four input points, but you cannot open the door to change your leads. Consequently, you mount four ice cube relays on an open DIN rail space. You run the 240–volt power supply input to the first relay, the 90–volt direct current output to the second relay, a clamp-on ammeter from the power supply input to the third relay, and a **temperature probe** to the fourth relay. You can now connect common leads from the relays to your meter. You also mount a selector switch in a spare knock-out hole on the panel door and connect it to each of the ice cube relays. You are using a four-position selector switch so that no two relays can be simultaneously energized. You are now ready to conduct your test.

You can change meter functions of the TX3 from the laptop outside of the panel. With the machinery running, you use the selector switch in conjunction with your laptop computer to monitor any one of six critical values: (1) the power supply input alternating current voltage; (2) the power supply input frequency (because the TX3 has a dual display, you can simultaneously read the frequency and an alternating current voltage value in the AC function); (3) the power supply input amperage; (4) the power supply output direct current voltage; (5) the power supply output AC+DC voltage, which identifies any alternating current components on the direct current output (again, the dual display simultaneously shows both the alternating current and the direct current values on the screen); and (6) the inside cabinet air temperature.

Initially, all test values appear to be normal; the AC input voltage, amperage, and frequency are steady. The direct current outputs are probably acceptable, but there is a measurable alternating current component that catches your attention. As the test continues, the alternating current input amperage begins to increase slightly. You also notice that the direct current voltage is not steady. Finally, you begin closely watching the AC+DC output voltage because you see increasing alternating current values and fluctuating direct current output. The AC+DC values continue to deteriorate until the clutch begins to slip. You also notice that the air temperature inside the cabinet is increasing.

You are certain that the direct current power supply to the clutch is faulty. There is no need for further testing, because the power supply is under warranty. After you have shut the equipment down, de-energized the panel, and opened the panel door, you carefully check the power supply for heat. It is excessively hot.

CHAPTER REVIEW

In spite of the availability of sophisticated solid-state test equipment, the multimeter as either a moving-coil volt-ohmmeter (VOM) or a solid-state digital multimeter (DMM) continues to be the basic test instrument for the industrial electrician. For effective and safe operation, however, the multimeter must be properly used. There are two essential precautions when using any multimeter:

1. The multimeter selector switch must be set for the correct test function and range before starting any circuit testing.
2. All resistance testing must be done on de-energized circuits.

The multimeter's most frequent testing function in industrial electrical work is the indication of the presence or absence of a voltage. The voltage test can be used to verify the conductive state of the circuit. Secondarily, the resistance-testing capabilities of the multimeter can be used for testing the continuity of a de-energized circuit.

In many instances, the multimeter is used to merely indicate the presence of a voltage or resistance. The actual value of that voltage or resistance is inconsequential to the purpose of the test.

The multimeter can be used effectively for the following resistance tests:

1. Phase-to-phase and phase-to-ground testing for the purpose of locating short-circuits
2. Continuity testing for the purpose of verifying completed circuits in electrical wiring or electrical devices
3. Lead or conductor identification for the purpose of labeling lead sets (pairs)
4. Capacitor testing for the purpose of identifying shorted or otherwise faulty capacitors
5. Diode and other solid-state device testing

Present DMM technology has added many features to the basic multimeter. Scale resolutions of 0.000 volt and 00.00 ohm are available. Meters will often have functions such as internal amperage ranges (usually limited to 10 amps or less) and external current probe connections (often allowing measurements as high as 500 amperes), internal temperature functions used with a **thermocouple** probe, dual displays of both direct current and alternating current voltage or alternating current voltage and frequency, capacitance measuring capability, resistance and diode testing capabilities.

Top-of-the-line meters are now available that can be directly linked with computers to enhance the graphic displays, to print the data in graphic and numerical formats, and to permanently store the information in computer memory.

THINKING THROUGH THE TEXT

1. What three areas are mentioned as necessary meter safety precautions when using a standard multimeter?

2. During a conductivity test on a live circuit, what value will the meter indicate across an open contact in the circuit?

3. During the same conductivity test on a live circuit, under what two conditions will the meter read zero volt?

4. What would a low-resistance phase-to-phase test on isolated conductors indicate? What would a low-resistance phase-to-ground test indicate?

5. Why is the indication of a low resistance value a satisfactory test for continuity?

6. When testing a diode with a multimeter, the leads are reversed and two readings are taken. Why does a high- and a low-resistance reading after the leads are reversed indicate a good diode?

7. What are the three primary advantages of using an electrical test meter linked to a computer?

CHAPTER 9

Specialized Tests and Equipment

OBJECTIVES

After completing this chapter, you should be able to:
- Select specialized test equipment that best suits your troubleshooting needs.
- Employ conventional values of voltage, amperage, and resistance in testing procedures to facilitate your unique troubleshooting requirements.
- Do a variety of specialized tests designed to reduce your troubleshooting time.
- Develop innovative applications for the test equipment you regularly employ.

NEW TROUBLESHOOTING POSSIBILITIES

In this chapter, on-line troubleshooting will be taken a step further. You will be shown testing techniques that are possible because of the solid-state equipment presently available. Troubleshooting accuracy and safety are in no way compromised—they are often enhanced. Through a careful use of troubleshooting techniques—in conjunction with some truly fine test instruments—you will find ways to dramatically reduce the amount of time spent in troubleshooting motors, electrical controls, and other industrial electrical equipment.

Test Instruments You Will Be Using

Most test equipment normally required for plant maintenance work is provided by the employer. Nonetheless, the majority of the equipment described in this chapter is within the budget of any electrician interested in making a prudent investment in professional tools. To give you a general idea of price range, a comparison of the union journeyman's hourly wage with the price of several of these instruments was made. The most expensive instruments are the hand-cranked megohmmeter and the Scopemeter. It would require approximately 31 hours of wages for the hand-cranked megohmmeter and approximately 45 hours of wages for the Scopemeter. On the other hand, the highest-quality digital clamp-on ammeter would require only about 9 hours of working time. The least-expensive digital clamp-on ammeter would require about 4 hours' work. At the low end of the price scale, some of the equipment described in this chapter would take the equivalent of no more than 2 or 3 hours' wages.

As you read through this chapter, you will need to be discriminating as to which of these test instruments are worthwhile for you. That decision will primarily be based on the frequency of their use and the cost of downtime in *your* application.

Finally, you should be aware that the following sections describe only a small portion of the test equipment currently available. The troubleshooting techniques in this chapter are a mere sampling of what you can do in the field to expedite your troubleshooting speed. *The important lesson of this chapter is the value of improvising. If you can learn to devise appropriate testing techniques in your situation with the test equipment you have, the purpose of this chapter will have been achieved.* Nonetheless, be careful that you understand the limitations of the equipment and the field conditions before conducting tests. Misuse of a meter can be hazardous as well as void your warranty.

Brand-name equipment is specifically identified in this chapter. It is not intended, however, to suggest that only the instruments named are suitable for the testing conditions described. As you shop for test equipment, you will find many other makes of instruments that will serve you well for your troubleshooting work. The information in this chapter is merely intended as a guideline for your instrument selection.

TROUBLESHOOTING WITH AMMETERS

The presence of a voltage on the terminal of an electrical component is not positive proof that the component is functioning. If the component itself is internally open, the voltage test will be inconclusive. On the other hand, if the component is open, the current draw will be zero. Thus, a current test will give additional information that is unavailable with only a voltage test.

The Advantage of Knowing a Current Value

If the current draw is abnormal, something is faulty in the electrical component. An explanation of an "abnormal" current draw follows later. For now, having added the current measurement, you now have three test values that can be used in your on-line troubleshooting work:

1. *The presence (or absence) of a voltage* at the electrical power component terminal. This is a test of the circuit up to the component, but tells you nothing about the condition of the component itself.

2. *The presence (or absence) of a current draw* on the electrical power component circuit. This test indicates that the component is electrically conductive, though it does not necessarily indicate whether or not it is operating normally.

Note: The voltage test must be done at the power component's terminal. The current test, however, can be done at any point along the circuit, since all parts of the circuit carry the full current value if there are no parallel legs. This can be used to your advantage, since a full circuit check can be done at any point on the circuit. For instance, the presence of a current on a single limit switch wire would indicate, with some qualification, that the circuit is operational.

3. *A specific current value* that can be used as a comparison with the current draw of other similar electrical circuits. This test compares the current value of the power component under test with that of a similar component. In many cases, this test will establish whether or not the component you are testing is functioning normally.

The following example will illustrate what has been said in the three points above. Say you are troubleshooting a live circuit line on the ladder diagram used elsewhere in the book. The circuit you are testing has a relay at the end of the line that is not closing. Consequently, you are measuring the voltage and current draw of the relay coil.

1. Your first test is a voltage test between the common grounded conductor and the wire to the relay coil. The multimeter reading is 120 volts. Therefore, you know the circuit to the relay coil itself is complete.

2. Your second test is done with a clamp-on ammeter. You get a reading of 0.29 ampere. You know that the coil is carrying a current, which means that it is not open. *That is, the coil itself is operational,*

but the mechanical part of the relay is not responding.

3. Your third test is on a similar relay coil circuit. It is only important that the second circuit, which you are using for comparison, is controlling a single relay identical to the actual circuit you are troubleshooting. In this case, the test result is 0.17 ampere. The reasons will be given later, but from this information you know that the relay drawing 0.29 ampere is not properly closing. Though the coil is operational, something is wrong with the mechanical part of the relay.

Though the above problem of a relay with a functional coil that is not mechanically closing is an infrequent type of failure, you will occasionally encounter it. In this case, if you were to shut down the system and examine the relay, you might find, for example, that the shading coil had broken, preventing the armature from seating. This could cause the relay contacts to remain open or seat poorly enough to cause premature burning and subsequent failure.

Evaluating the Procedure

In this testing sequence, you determined that the relay had a voltage across it, which meant that it was drawing current. Normally, that would indicate an operative circuit. In this case, however, you were able to establish with additional ammeter values that the relay was not closing. By comparing the current draw of a similar relay with the one under test, you determined that the relay, which proved to be defective, was drawing too high a current.

A solenoid or relay, which are both electromechanical, will draw excessively high currents if the path for the magnetic flux is broken by separating the steel. If the armature (the moving steel) cannot move to its fully closed position, there will be an air gap, and the coil current will significantly increase. The precise amount of increase in amperage values is not the concern. By merely comparing the power component with a similar component in the circuit, it is possible to determine that the current draw is excessive.

Note: Interestingly, a broken or displaced shading coil will not cause an increase in current as long as the armature properly seats. The armature will be noisy, but the current will not be affected since the magnetic saturation of the steel has not changed with the loss of the shading coil winding.

In Chapter 8, you learned how to read resistance to verify two parallel power components in a single circuit where the combined resistance of two identical electrical components is half the value of either of the components measured individually. A clamp-on ammeter capable of reading low current values can perform the same function if the circuit is energized. In this case, however, the current values are added. That is, two parallel relays—each drawing 0.17 ampere—will draw 0.34 ampere on a single wire feeding both relays. Thus, you could test a parallel circuit in a similar manner. You would first take a current reading on a single power component that was identical to the components on the parallel circuit. In the final test on the parallel circuit, if the current value is double the value of the comparison circuit, you know that both components are functioning. If, however, the value is equal to the comparison circuit, you know that one of the parallel components is open.

Ammeter Reading Ranges

Before giving you an example of an ammeter used in testing a hydraulic solenoid circuit, we need to give you some specific information concerning clamp-on ammeters. Did you notice the current value assigned to the relay coil in the previous example? It was 0.17 ampere. In fact, this is the approximate current draw of an actual **NEMA** size 1 contactor operating at 120 volts AC.

For many of the testing procedures described, you will need to measure values in tenths and hundredths of an ampere. With a digital ammeter capable of displaying two digits to the right of the decimal point, this precision is attainable. Figure 9–1 shows two manufacturers' ammeters that are capable of this degree of precision.

Generally, however, this degree of precision is below the reading range of most moving-coil clamp-on ammeters. In all practicality, moving-coil meters could not be used for this testing.

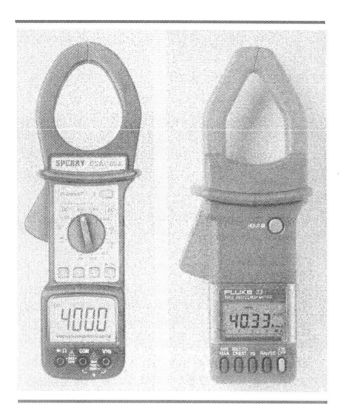

Figure 9-1 Two high-quality digital clamp-on ammeters. The A. W. Sperry DSA-1003 meter on the left is a combination AC and DC meter. The FLUKE 33 is on the right. Both meters display true RMS values and have functions such as Peak Capture, Maxmum/Minimum, Range Selection, and Hold. *Courtesy of A. W. Sperry Instruments, Inc. and FLUKE Corporation.*

Note: You can increase the sensitivity of any ammeter by wrapping the conductor around the jaws. For example, if you wrapped the conductor around the jaws twice, you would double the reading on the scale. If you made a jumper lead that was wrapped around the ammeter's jaws ten times, you would increase the reading by 10. With ten turns, a 0.8–ampere measurement could be taken on a 15–ampere scale since it would read as 8 amperes. In this way, analog (moving-coil) meters can be used for lower-range measurements. The impracticality of this method is the effort required to add jumper wires to existing circuits in order to wrap them around an ammeter's jaws.

Because of the sensitivity of digital ammeters when they are operating in low ranges, there is a necessary precaution you must take to avoid inaccurate readings. The stray magnetic fields, or EMI, in electrical panels can generate a reading on the meter. You will find that meter position becomes a variable. With no wire in the meter jaws, if you hold the jaws directly in front of an energized coil, you may read 0.00 ampere. By moving the meter to either side of the coil, the amperage reading will increase. You obtain the zero reading when the jaws are in a position where the alternating current circuit is balanced. It is much like the zero reading you obtain when you clamp the ammeter jaws around both wires of an alternating current circuit. Before putting the wire through the jaw of the ammeter for a test requiring precise values, it is necessary to hold the meter in the exact position of the test to determine if there is a value on the meter readout from the EMI. The most accurate testing is done by keeping the meter as far as possible from these magnetic fields.

Actual Ammeter Testing

You will again use the mold-close circuit from Chapter 5. Figure 9–2 is line 19. The final wire in the circuit is wire 40, which is connected to the mold-close solenoid (solenoid F). In the example you are about to study, you have a similar problem to the one you tested in Chapter 5, because the mold will not close. You first test line 19 for a control voltage. In this test, however, you find that there is a control voltage to the solenoid itself. Thus, your testing will be confined to the solenoid to determine the fault. The machine is stalled on the nonfunctioning mold-close cycle.

Voltage Verification Test

Your first test will verify the voltage to the solenoid coil. You will use a voltmeter to verify the presence of full control voltage. The test indicates 120 volts on the solenoid lead wire 40. However, the presence of a voltage at the solenoid terminal does not indicate that the solenoid coil is functional. The coil could be faulty. For the sake of this example, the coil is operational.

Current Test

Test purpose. To verify that the solenoid coil is operational.

ampere. Because this current draw is less than the current draw from the previous test of 1.4 amperes, you know that solenoid F (mold close) is electrically operational, but is mechanically faulty.

You learned earlier that when a relay armature fails to completely close, the current draw will be excessively high. The same is true of a solenoid valve. In the testing example that was just given, you saw a solenoid valve rated at 0.8 ampere drawing 1.4 amperes. You could conclude from this that there was a mechanical problem in the valve that was preventing it from shifting. Whatever the mechanical problem would eventually prove to be, it is safe to go directly to the valve with no more electrical testing.

There is no current formula for current draw that will indicate the degree of the problem. This type of testing is merely done to compare a known normal value with an unknown value. Nonetheless, even though values cannot be assigned for the current, the testing procedure is extremely reliable and can save you a great deal of time in locating mechanical problems from the electrical panel.

A mechanically jammed **spool valve** would well be in the range where you would expect to find a very high current deviation from normal. However, you would not find a deviation if the spool or armature was jammed in the fully closed position. Do not try to take this information as the basis for test formulas. If you find a current variation between identical solenoid (or relay) coils of more than 10 or 15 percent, it merits further verification of the mechanical condition of the component.

Note: A high current on a solenoid valve could also be the result of turn-to-turn shorts in the coil itself. Shorts will produce a hot coil. If the coil is not excessively hot, this possibility could be tentatively checked by comparing the resistance of two identical coils.

Other Ammeter Tests

Other on-line troubleshooting can be done when you have the capability of measuring these small current values. Complete circuits can be verified as operational by taking an ammeter reading at any point on that circuit. In other words, you can reduce your entire testing sequence to one test point in order to verify an entire ladder diagram circuit line *if the circuit is properly functioning*. By clamping the jaws of the meter around a limit switch wire, you could verify that the entire circuit was operational. If there is no current reading, however, you have no indication of where the break is and will need to rely on other troubleshooting techniques.

The greatest difficulty with this testing procedure is the physical size of the ammeter jaws. It is often difficult to clamp the wire you want to test in the limited space available. Nonetheless, the concept of this testing procedure may well open very simple testing methods to your specific applications. The FLUKE T5-600 or T5-1000 shown in Figure 9–4 allows ammeter use in highly confined spaces. However, the T5 will only read tenths of an ampere (0.1), which is above the current draw of small relays and solenoids.

Most better-quality digital clamp-on meters have a "peak" hold function. There are a number of test procedures that can be done with a peak current

> **A DAY AT THE PLANT**
>
> *For the sake of interest, let me give you some values that I obtained from bench-testing an actual hydraulic solenoid valve. I was using a coil that was rated at 0.64 ampere. The total spool travel of the valve was 0.125 inch. Using a clamp-on ammeter capable of reading hundredths of an amp (0.00 ampere) I could read a tenth of an ampere variation by moving the valve spool 0.002 to 0.003 inch off its seat from the fully closed position. Moving the valve spool 0.020 inch off its seat resulted in a current increase of 20 percent. At 0.05 inch off its seat, there was a full 100 percent current increase. The actual measurements and percentages are not significant in this example. What is significant is that the meter can detect such a small initial variation.*

120 CHAPTER 9 Specialized Tests and Equipment

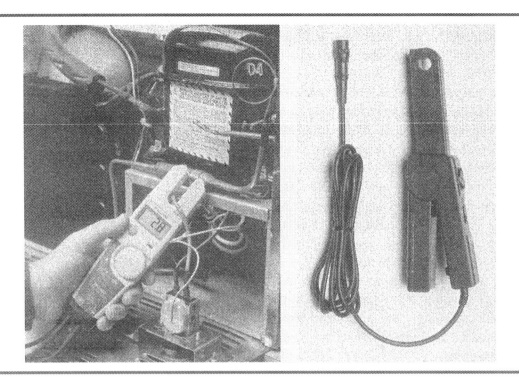

Figure 9-4 There are specialized ammeters or accessories that will work effectively in confined electrical panel spaces. The flank T5-600 or T5-1000 (show on the left) use a small, open jaw that can be used in limited space. FLUKE and other manufacturers often supply small ammeters clamps that are used with a DMM. FLUKE's 80i-110s Current Clamp is show on the right. *Courtesy of FLUKE Corporation.*

value. It is useful in determining motor starting current. Starting current values can often provide useful troubleshooting information when there are mechanical or electrical overload conditions. In normal motor start or run conditions, peak current values can be used to size fuses to maximum operating conditions. The peak can also be used as a comparison with normal currents if you are trying to locate a random surge. The clamp-on meter can be left hanging on the conductor under test in the peak setting. If the equipment goes down because of a high current draw, the meter will record and hold the information until it is reset. For example, suppose you take a reading on a circuit and find that it is normally operating at 16 amperes, even though the 20–ampere circuit-breaker is occasionally tripping. After hanging an ammeter on the conductor, you find a 24–amp "peak" reading on the meter when the circuit-breaker trips. You now know that some form of intermittent overcurrent condition exists that must be corrected.

In addition to voltage and resistance functions, which are measured with leads from a multifunction clamp-on ammeter, solid-state ammeters often provide other functions as well. Many manufacturers offer true RMS meters that are useful in locating harmonics. Some of these meters simultaneously display both frequency and current. Better-quality meters will have a soft mode that reduces fluctuations in readings by displaying a 3–second running average. Most better clamp-on meters will record peak, minimum, and average values of RMS current, softened RMS, and frequency. Some will measure either alternating current or direct current displayed with a resolution to hundredths of an ampere on their lowest range.

Electrical malfunctions usually represent one of three broad categories: (1) the mechanical failure of a component, such as the failure of the internal mechanism of a limit switch, (2) deterioration of electrical contact elements, such as the excessive pitting and burning of relay contacts, and finally, (3) insulation failure, such as the internal grounding of

A DAY AT THE PLANT

*In a field test I conducted, the 3-phase, 480–volt, 400–ampere main circuit-breaker on a pipe-forming machine tripped during production for no apparent reason. After an inspection of the machine, we put it back into operation. In the next few weeks it tripped two or three more times. I had reason to believe that the circuit-breaker itself was failing because it had been used as the daily power shutdown for years. However, before buying an expensive circuit-breaker, I needed to be certain that there was no electrical problem behind the tripping, because it occurred when the machine was under its heaviest loads. During the next run of large pipe, the circuit-breaker again tripped early in the day's production run. At that point, I put two digital ammeters on two phases of the incoming power in the 600–ampere **branch circuit** disconnect. After first monitoring the current draw while the machine was in full production, I set both ammeters on their "peak" functions. Several hours later, the circuit-breaker again tripped.*

This time I had my answer. Both meters showed a peak current value substantially under the 400–ampere rating of the circuit-breaker. It was clear that the circuit-breaker itself was mechanically failing. The replacement circuit-breaker eliminated the problem.

Why two peak meters rather than one? Because with two meters I would cover any heavy single-phase loads as well as the normal three-phase loads. Though direct, momentary shorting to ground was unlikely, however, I still risked missing a momentary phase-to-ground fault on the phase without the ammeter.

motor windings. The majority of the troubleshooting procedures described in the early chapters of this book deal with component failures of some sort that are attributable to mechanical failures or deterioration of electrical contact elements. The third cause of electrical malfunctions is insulation failure. Insulation failure and insulation testing is the subject of this section.

Electric Motors

Electric motors are probably the most important and frequently used end users of electrical power within industrial settings. Therefore, motor testing procedures are important in the overall troubleshooting program of any industrial plant.

A motor that has failed will most frequently have an insulation-related problem. Whatever condition outside of the motor initially caused the problem, the resultant motor winding heat will destroy the insulation, which is the immediate cause of motor failure.

A motor that has completely failed is generally easy to locate with a standard ohmmeter. If the insulation is charred, an ohmmeter reading to ground will show very low or zero resistance. A reading between leads may also indicate that they are open by indicating infinity (∞). When the motor has completely failed, your nose is as effective as any meter in your toolbox. You just use the meter to sound "scientific" when explaining a $400 motor replacement.

Since the subject of this book is fast troubleshooting, look at a simple procedure that can be used when verifying a totally damaged motor. This motor will be completely shorted to ground. A motor that is in the early stages of insulation failure will require testing with a megohmmeter, which will be discussed later. Rather than going to the motor itself and disconnecting wire nuts or terminal screws, it is much faster to open the motor controller, making certain that all necessary lockout procedures have been followed, and test from phase-to-phase and phase-to-ground with an ohmmeter on the motor side of the controller. This can result in significant time savings when the motor is buried deep in the machinery. If the test indicates very low resistance to ground or open windings, then your time is well spent in starting to work on the motor itself. If, on the other hand, the tests do not indicate abnormal conditions, you may want to look elsewhere before going to the effort of opening motor cover plates or disassembling the motor for a visual check. Do not use a megohmmeter for this testing until you understand the caution of its use near solid-state equipment. This is discussed later in the chapter.

You can check the condition of a motor just as accurately with an ohmmeter lead on the motor's supply line at the motor's starter as you can with the ohmmeter lead on the motor's wire terminal inside the motor's electrical cover plate. Until you need to visually examine the interior of the motor, you have gained little by going to the effort of removing cover plates from the motor's terminal box. If the motor is inaccessible, you will save time by doing the testing from the most accessible point in the system. That will be from a Motor Control Center (MCC), a motor controller, or a lockout disconnect by the motor.

Make certain that you have completely isolated the motor circuit with appropriate lockout procedures before doing any motor testing as suggested above.

Insulation Materials and Failure

Before finishing this section with a description of megohmmeters and their use, take a quick look at insulation materials. Understanding what they are will help you understand what is happening when they fail.

Modern electric motors are wound with "magnet wire." Magnet wire includes a broad range of round and rectangular conductors used in coil windings for motors and transformers. Magnet wire has an enamel insulation coating. In fact, the term "enamel" (or "varnish") may refer to any one of approximately thirty different insulation materials. Each of these materials has some advantage, whether it is in lower-cost, higher-temperature range stability, greater flexibility for handling during winding, or any of a number of other specialized characteristics. A chart of enamels includes such material names as acrylic, nylon, polyvinyl formal, epoxy, polyesters, and even ceramic materials. General-purpose enamels tolerate temperatures up to 105°C (or 221°F). Special-purpose enamels extend the range to higher limits.

Today's high-efficiency motors operate at elevated temperatures that demand even better grades of wire insulation. In addition, more alternating current motors are coming off the assembly line with inverter-grade insulation, even though they are not sold as inverter-rated motors. Inverters, which are alternating current motor speed controls, can produce high-voltage transients that greatly exceed the line voltage rating for the motor. Unless the insulation is able to tolerate these high voltages, the insulation will be compromised and will then quickly fail.

There are other insulating materials besides the wire enamels used in general-purpose motor and transformer windings. After the windings are in place in the motor frame, they are often coated with baking varnishes or potted in epoxies. Oil-impregnated papers, fiber materials, and various plastics and rubbers are used as separating materials between coil groups or lead conductors.

What causes a motor to burn out? With the exception of grossly excessive current loads, which actually cause the conductors to melt, almost all motor losses are a result of some form of insulation failure. Whether it is a progressive insulation deterioration over a period of years or rapid deterioration because of moisture or high current loads, the motor finally burns out because the insulation can no longer isolate the voltage differences in adjacent wires or coils.

Without trying to be highly technical, the reasons behind insulation failure can be classified into four broad categories:

1. *Heat-related breakdown.* This can be a long-term condition where organic materials in the insulation material degenerate to a point at which they can no longer provide the insulation values necessary for the application. It can also be caused by short-term conditions in which the insulation breaks down rapidly. A motor operating at 15 or 20 percent overload conditions with oversized thermal protection will quickly exceed acceptable temperature ranges, and may burn

out in a matter of hours—or even minutes—of operation.

2. *Moisture-related breakdown.* Moisture is a major cause of insulation failure. Small droplets of water from condensation on the motor windings will cause arcing between conductors when the motor is restarted. At the point of the arcing, the organic materials will be carbonized as they are burned. This, in turn, becomes a conductive bridge between the windings, which will cause further arcing and heat.

Note: Properly installed motors that are subject to repeated contact with moisture will use a coil heater to keep the motor warm during shutdown. If a motor's insulation is damp—but otherwise unharmed—it can be dried in place. This is best done with external heat sources, such as hot air or heat lamps, which are directed on the motor. However, excessive heat must be avoided.

3. *Mechanical-related breakdown.* Insulation can be destroyed by mechanical problems such as vibration or physical contact with moving parts. A motor could burn out if a worn bearing allowed the rotor to touch the stator. The frictional heat would cause a hot spot on the stator, resulting in a localized winding burnout.

4. *Oil- or chemical-related breakdown.* Oil or chemicals drawn into motor windings can destroy insulation. The motor installation area must be kept free of foreign chemicals and the motor itself must be properly lubricated to avoid overflow into the winding area. Non-detergent oils are specified for motor sleeve bearings. Detergent oils have a greater ability to wick from the bearing to the shaft where it can be slung on the exposed windings.

Megohmmeter Insulation Testing

In each of the three causes of insulation failure given above, there will usually be early warning signs. The test instrument used to measure the insulation value of a winding is called a megohmmeter. Its name comes from its **meg-** (million) ohm resistance reading range. This instrument is used to test the resistance value of a winding under a load condition. The value read on the megohmmeter is the insulation breakdown point given in ohms of resistance. The scale is in megohms. Refer to Figure 9–5 for an example of an analog megohmmeter scale. Therefore, a reading of 50 on the scale means that at the test voltage produced by the meter, the insulation offers 50 million **ohms** of resistance. Taken from Biddle Instrument's registered trade symbol Megger®, the megohmmeter is often called a "megger" in the trades.

It is not possible to give absolute values for "good" and "bad" insulation. Insulation resistance values should be considered relative. Values could vary considerably for one motor or machine tested on three consecutive days without indicating poor insulation. It is the trend in readings over a time period that are the most significant. Periodic readings that show decreasing resistance indicate future problems. Periodic testing is, therefore, your best approach to preventive maintenance of electrical equipment. Because megohmmeter tests are temperature dependent, you should make tests at about the same temperature, or correct them to the same temperature.

A new industrial motor, or one that has been reinsulated and baked by a rewinding shop, will show close to infinite (∞) resistance between the coils and ground. However, an industrial motor in service with good, dry insulation will show some resistance value from phase to ground. Maintenance professionals often use the 1–megohm rule to establish the allowable lower limit for insulation resistance. By this standard, insulation resistance should

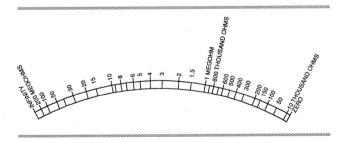

Figure 9-5 A typical a analog megohmmeter scale. Notice the scale divisions on the low-registration of portion of the scale form 10,000 to 800,000 ohms. Scale division on the high-resistance portion of the scale are from 1 megohm (1,000,000 ohms) to 200 megohm (200,000,000 ohms). *Courtesy of AVO International/Biddle Instruments.*

be approximately 1 megohm for each 1,000 volts of operating voltage, with a minimum value of 1 megohm. For example, a motor rated at 2,400 volts should have a minimum insulation resistance of 2.4 megohms. In practice, megohm readings normally are considerably above this minimum value when insulation is in good condition.

By taking readings periodically and recording them, you have a better basis of judging the actual insulation condition. Any persistent downward trend is usually fair warning of impending trouble, even though the reading may be higher than the suggested minimum safe values. Equally true, as long as your periodic readings are consistent, they may be acceptable, even though they are lower than the recommended minimum values.

Figure 9-6 shows two types of megohmmeters: battery powered and generator (hand-crank) powered. Either type is equally effective at a given voltage range. In both cases, full-rated voltage is imposed on the windings being tested. The megohmmeter leads are used to energize the motor coil with high-voltage, low-current power. The simplest megohmmeters to use are battery powered. Battery-powered meters are available from low-voltage output to 1,000–volt output models. The higher-voltage models are generally the most expensive. Megohmmeters use two leads for the insulation value test. A third "guard" lead is sometimes used to correct the error of current leakage *over* the surface of the insulation. Analog megohmmeters are operated and read much like a standard VOM multimeter. Digital megohmmeters are read like a DMM.

There was a section in Chapter 8 describing computer-enhanced multimeters. The same provision is also available in some megohmmeters. With a megohmmeter interface option, you can both enhance the readout information during a test, and have the ability to download information to computer memory. As you saw in the section on computer-enhanced multimeters, downloading graphic information from a computer-enhanced megohmmeter is a great aid in permanently storing test data for future trend analysis of motors throughout a plant.

During megohmmeter testing, a motor winding can act as a capacitor because it has conductors separated by insulation. After any insulation test, motor or transformer leads should be shorted to ground to dissipate this charge before working on the leads. This is especially true with either large or high-voltage motors.

As you can see in the megohmmeter examples, varying test voltages are used for each testing condition. Though some megohmmeters use alternating current voltage output, most hand-crank and battery-operated testers for general use have a direct current voltage output. The test voltage is usually slightly above the operating voltage of the equipment under test. AVO International/Biddle Instruments recommends the voltage selections given in Table 9–1.

There is a necessary caution when using a megohmmeter. Because general-use meters actually

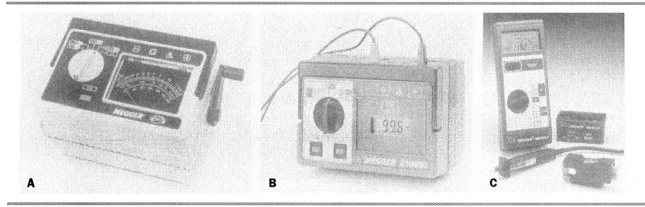

Figure 9-6 Three typical megohmmeters are shown in this figure: (A) A hand-crank analog megohmmeter with multiple voltages to 1,000 volts. (B) A digital battery powered megohmmeter with similar functions as the along meter in the first illustration. (C) A combination megohmmeter and DMM with both digital and simulated along display. *Courtesy of AVO International/Biddle® Instruments.*

Table 9-1 DC test Voltage selection on a megohmmeter is chosen on the be basis of the operating voltage of the equipment under test. *Courtesy of AVO International/Biddle® Instruments.*

Equipment AC Rating	DC Test Voltage
up to 100 volts	100 and 250 volts
440 to 550 volts	500 and 1,000 volts
2,400 volts	1,000 to 2,500 volts or higher
4,160 volts and above	1,000 to 5,000 volts or higher

have an output of 500 or 1,000 volts, it must be used carefully. Never touch a "hot" megohmmeter lead. Believe the manufacturers' directions when they say, "Don't do it!" When used at the voltage setting of the motor under test, modern direct current megohmmeters are nondestructive, which means they do not damage the motor windings during the test. However, there are certain kinds of tests for which the meter must never be used. Though you may test the windings of a single-phase motor, you must never have a capacitor (for a capacitor-start or capacitor-run motor) in the circuit while you are testing. The high voltage will puncture the capacitor. The same is true for solid-state equipment. *Never use a megohmmeter if it might come in contact with solid-state circuits.* If you are testing a motor powered from a variable speed drive, *make certain that the motor is completely isolated from any wiring to the drive.* The same caution applies to **temperature detectors** embedded in motor windings and motor-mounted thermal switches.

Though generally nondestructive, do not use unnecessarily high test voltages. Follow the values given in Table 9–1. If you are testing a motor that has been wet, always do the initial testing on the lowest voltage settings. Do not use higher voltage settings if the windings show abnormally low resistance. Dry the windings before doing further high-voltage testing.

Also remember that resistance values must be temperature compensated. For every 10°C increase in temperature, halve the resistance, or for every 10°C decrease, double the resistance. For example, a 2–megohm resistance at 20°C reduces to ½ megohm at 40°C. Compensated resistance values may be determined from the nomograph given in Figure 9–7. The temperature is that of the actual coils in the motor or appliance during test.

A DAY AT THE PLANT

Can a megohmmeter actually damage insulation? To find out, I deliberately tested a new 115–volt universal motor with a 1,000–volt battery-powered megohmmeter. At the beginning of the test, the meter showed almost 6 megohms of resistance. I held my finger on the test button for less than 15 seconds and then observed the meter display counting down to a very low value. After letting the motor rest for several days, I verified the results with a generator megohmmeter set on the proper voltage range of 100 megohms. The latter test showed that the motor had a dead short to ground. I had totally destroyed a light-duty 115–volt motor with a 1,000 megohm test. The lesson learned is that even with a battery-powered megohmmeter, the voltage output is exactly what its rating says it is. An industrial 480–volt motor may not be damaged on the 1,000 megohm setting. However, the light-duty insulation of this universal motor was unable to tolerate the high-voltage stress.

With this background in insulation and the function of a megohmmeter, you should be able to see this test instrument's potential. You could use it much like the resistance testers described earlier in this section. Now, however, by using the megohmmeter, you can do an actual resistance value test from the motor controller or motor disconnect.

Note: If all wiring is in good condition, a test from the controller will be adequate until final testing is done on the motor itself. However, if any part of the branch circuit is failing, the megohmmeter reading will show a low resistance value irrespective of the condition of the motor. Conversely, if the wire is open (but not grounded) between the controller and the motor, the resistance

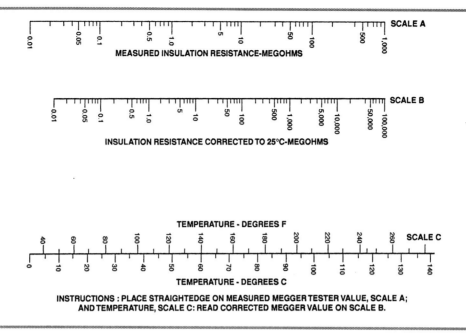

Figure 9-7 This nomograph can be used to correct winding temperatures to 25°C. The correction is based on rotating machinery with Class B insulation. *Courtesy of AVO International/Biddle® Instruments.*

value would be high (approaching infinity) irrespective of the condition of the motor. Consequently, the first step is to verify that there is resistance between each of the motor leads that approximates the expected values of the motor coils themselves. Subsequently, if a low resistance is encountered, the branch circuit must be isolated from the motor in order to identify which part of the circuit is actually showing leakage to ground.

A plant refrigeration unit developed a leak, which required a service call. Before the leak was discovered, there was indication that the compressor had been running hot. Several weeks after the leak was repaired, the compressor blew a fuse and was shut down. This particular refrigeration unit had two circulation pumps powered from the main electrical circuit and one hermetically sealed compressor on a thermostatically controlled circuit with a starting relay. It was a three-phase, 240-volt unit. The following tests were done:

1. A standard ohmmeter was used for the first testing. All three legs of the three-phase power-to-ground branch circuit tested as open, giving no indications of direct-ground shorts. Secondly, both the circulation pump circuit and the compressor circuit were tested. The phase-to-phase circuits showed continuity across each set of motor windings, indicating that no windings were open. Additionally, each motor tested as being open from phase to ground.

2. A second group of phase-to-ground tests was done with a megohmmeter from the same test points. The TIF IT990 Electronic Insulation Tester shown in Figure 9-8 was used. This is a battery-operated tester rated at 500 volts. The second test showed a reading of approximately 5 megohms from phase to ground on the circulating pump circuit. On the other hand, a reading of about 300 kilohms was shown across the compressor terminals to ground. Since this last reading was less than the acceptable limit, there was strong evidence that the compressor was failing.

3. In order to verify the conclusion, the fuses were replaced and the compressor control switch opened. With only the circulation pumps running, the system operated normally. A clamp-on ammeter verified the expected current draw. The compressor was then turned on and immediately blew another fuse. Since the previous tests had shown that there was no ground-fault, it was ob-

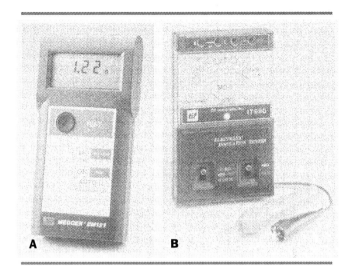

Figure 9-8 Battery-powered megohmmeters are also available at lower cost and smaller size. These meters can be conveniently carried in an electrician's toolbox. (A) The Biddle 120 series meters are available in either 500- or 1000-volt outputs. *Courtesy of AVO International/Biddle® Instruments.* (B) The Tif It990 has a 500-volt output. *Courtesy of Tif Instruments, Inc.*

vious that the compressor was drawing excessive current, which was evidence that the compressor winding insulation was, in fact, failing. During the compressor test, a digital ammeter was clamped to the circuit. The ammeter was set on the peak current function, giving the current value when the fuse failed. Again, the current values confirmed an internally shorted compressor motor.

Since the two circulation pumps were in parallel with the main power supply, the electrician only needed one test point for both pumps. However, since the main power supply came to the compressor starter relay, all of the testing was done from a single point. The first test was from ground to each of the three-phase legs on the line side of the compressor relay. This single test provided a megohm test on the two pump motors, since they were in parallel. Finally, a test from ground to each of the three-phase legs on the compressor side of the starter relay provided a megohm test on the compressor. No wires were disconnected for the entire test.

The purpose behind this explanation is to illustrate troubleshooting speed. By carefully choosing test points, and by using an appropriate piece of test equipment, the electrician was able to quickly and accurately locate the cause of the problem. There was no need to go directly to the motor terminals for the initial test. As a final verification, of course, the test was repeated on the disconnected motor leads. A megohmmeter is an excellent test meter for hermetic compressors because it is impossible to visually assess any part of the motor. For the same reason, however, it becomes a time-saving troubleshooting meter for any motor. Why visually check the motor by removing cover plates when it can be done more quickly from the magnetic starter?

Testing for Shorts with a Megohmmeter

A megohmmeter can be used effectively for other types of ground-fault testing as well. This can be demonstrated with a field example.

During the morning start of a production run, a conveyor lost its 120–volt control transformer fuse. The electrician checked the control system without finding a problem. The fuse was replaced and the conveyor was put into production. Several days later, the same blown-fuse sequence reoccurred. Again, the electrician could find no apparent problem. When the fuse blew a third time, it was time to locate the problem!

It was apparent that the cause was an intermittent problem that was not evident during visual inspection. Fortunately, the circuit was relatively simple. It consisted of a controller mounted on the wall. The motor drive was mounted in a pit under the conveyor, and several limit switches were attached to the conveyor track. There was an operator's station about 10 feet from the conveyor. After initial testing, the electrician determined that the controller was probably not at fault. The electrician was also fairly certain that the operator's panel was not shorting. But the electrician had reason to be suspicious of the limit switches, because the conduit was partially underground, and the limit switches themselves had cord connections to allow adjustment. In addition, the limit switches were in a potentially wet area.

Because a megohmmeter has a high-voltage output, it could not be used on a connected circuit. Consequently, each limit switch wire was disconnected from the power panel's terminal block and individually meggered. The conductors to one limit switch showed a very low resistance. Opening the limit switch cover clearly showed what was happening. The limit switch housing was filled with production sediment. At the end of each day's run, operators cleaned adjacent production areas with high-pressure water hoses. A careless operator was occasionally splashing the area by the limit switch. Condensation overnight was wetting the sediment inside the switch and causing a problem the next morning.

Meggering the limit switch circuits was the ideal way to locate the short.

Remember, however, that because of its high output voltage, a megohmmeter must be used with much more care than a DMM. This type of testing should always be done on the megohmmeter's low-voltage ranges; preferably using the 250 megohm range if it is available. It is mandatory that the testing be isolated from all circuits that might be damaged by high voltages.

A compact battery-operated megohmmeter, such as Biddle Instrument's 120 Series shown in Figure 9–8, is a relatively inexpensive unit that can be carried in an electrician's toolbox. These smaller insulation testers readily lend themselves to field use as described in the example above, though they often do not offer multiple test voltages.

TROUBLESHOOTING CAPACITOR PROBLEMS

In Chapter 8, you were shown how to use a VOM to test capacitors for internal shorts. Proper capacitor testing, however, can be more complex than finding a simple short, though internal shorts are frequent faults that produce a completely worthless capacitor. Before a capacitor shorts, it will usually start to leak internally. Most high-quality DMMs include a capacitor test function. However, there are also specialized capacitor testers that indicate internal leaking conditions. Refer to Figure 9–9 for an example of a typical capacitor tester.

Capacitor Use

Before capacitor testing may be discussed, you will need to understand the capacitor itself. Generally,

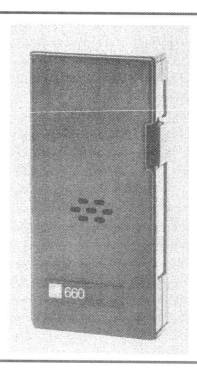

Figure 9-9 The TIF660 is a specialized Capacitor tester that audibly indicates a capacitor's condition. An approximate capacitance value can be Determined by timing the instrument's tone duration during the capacitor's changing cycle. This type of test instrument can be used when Specific digital values are not needed. *Courtesy of Tif Instruments, Inc.*

you will find four uses of capacitors in industrial applications:

1. They are extensively used as starting capacitors on single-phase motors. Since single-phase motors in industrial plants are primarily fractional (less than 1 horsepower), capacitor use will be confined to small pumps, blowers, and the like.

2. A second use of capacitors is for **power factor** correction. Power factor is the ratio of the actual power of an alternating current circuit to the apparent power. Inductive loads from motors distort the power factor balance. The power factor balance can be corrected with capacitance. Capacitance produces a leading power factor and is often used to correct lagging power factor where large inductive (motor) loads are encountered.

3. Capacitors are used in direct current applications. There will be power-conditioning capacitors in any circuit rectifying alternating to direct current.

4. Finally, you will encounter capacitors in both alternating current and direct current variable speed motor drives.

Caution is always in order when working with capacitors since they can hold a voltage potential (a charge) even though the capacitor is disconnected. With even a small capacitor, the charge may be sufficient to damage a meter during testing. With large motor drive and power-factor-correcting capacitors, the charge can be lethal. Always drain the disconnected capacitor by placing a resistance across the terminals for a specified period of time before handling or testing. Article 460 in the *National Electrical Code®* (*NEC®*) requires that capacitors will have a means of draining the charge. The voltage should drop to 50 volts within one minute after being disconnected. But do not take chances. An old-style capacitor may not have the bleed resistor.

The established unit for capacitance is *farads*. The farad is named in honor of the British scientist Michael Faraday (1791–1867). Since, however, the farad is a large unit, general-use capacitors are rated in **microfarads**, which is one-millionth of a farad. You will see this abbreviated as "mfd," "mf," or "µF."

Capacitor Construction

A basic capacitor is constructed with two conductive (metal) plates separated by an insulator. The separator is called the **dielectric**. In use, positive electrons will accumulate on one plate while negative electrons accumulate on the other. The ability of the capacitor to hold a charge is based on two factors: the size in surface area of the plates and the thickness of the dielectric. The thinner the dielectric, the greater the capacity. However, the breakdown limit of the capacitor is determined by the dielectric—or insulation—strength of the dielectric. Thus, dielectric material needs to be thin with a high insulation value. In most cases, the plates are made of foil with a paper or plastic dielectric. This allows the capacitor to be rolled and sealed in a can.

The basic capacitor used in alternating current applications is a form called an **electrolytic capacitor**. An electrolytic capacitor is a modification that uses a single plate as one conductor and a chemical compound called an **electrolyte** for the other.

The dielectric is a thin insulating film of oxide on the metal plate. Since the electrolytic capacitor is a direct current device, the use of these capacitors for alternating current circuits requires an additional modification. An alternating current electrolytic capacitor has two plates (anodes) that are common to one electrolyte. Thus, in operation, the two anodes are functioning alternately, which allows the capacitor to pass the alternating current. Figure 9–10 shows the construction of a typical capacitor.

Capacitor Failure

How capacitors are built was explained because that is the simplest way to understand why they fail. A capacitor most often fails because the dielectric (insulation) is no longer doing its job. If the plates (whether plates alone, or a plate and the electrolyte) touch each other, the capacitor will internally short and cannot hold a charge. There are generally two things that will cause a capacitor to fail:

Physical Damage

If the capacitor can is crushed, or if the terminals are twisted or moved, the physical dislocation of the foil and dielectric may either reduce the dielectric strength or actually cause contact between the plates. Vibration in service is another leading cause of failure for the same reasons. In any of these cases, the capacitor becomes worthless.

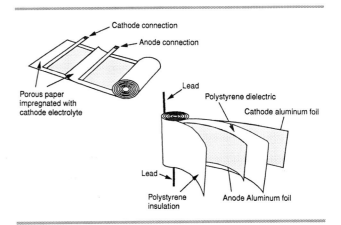

Figure 9-1 Typical capacitor construction. Construction the drawing on the left show a live show a basic electricity capacitor. The drawing on the right show a polystyrene capacitor construction.

Electrical Damage

If the charge voltage on the opposing plates exceeds the insulation value of the dielectric, a spark will travel through the insulation to the other plate. The effect will be an actual hole in the insulation material. At the very least, this breakdown in the insulation will reduce the ability of the capacitor to hold a charge. This condition is identified as a "leaky" capacitor, which means it will not hold a full charge. The electrons will "leak" from one plate to the other. If conditions are severe enough, the spark may actually weld the two plates together. This condition is identified as a "shorted" capacitor. In the first case, a multimeter test might not indicate a problem since there is no actual continuity between the plates. The only totally accurate test is to load the capacitor to its working voltage and measure its ability to hold the charge.

There are two other conditions that you may encounter in capacitor testing. One is an "open" capacitor. For some reason, whether through physical abuse or electrical burning of the internal lead connections, the leads are not making contact with the plates. The other is a "grounded" capacitor. In this case, a lead or plate has come into electrical contact with the metal cover. In either case, these capacitors are ready for the trash can.

Capacitor Testing

You can set up a workbench test for a capacitor by measuring electrical values and the loss rates. However, this is not the kind of field testing that is either practical or fast. If you frequently work with capacitors—and you do not have a capacitor function on your DMM—then specialized test equipment is in order. A number of manufacturers supply relatively inexpensive equipment for this purpose. TIF Instrument Company supplies a convenient handheld capacitor tester, TIF660, which is shown in Figure 9–9. As is so often the case, this capacitor tester can be used for other non-capacitor test functions. The TIF660 can equally be used as a continuity or diode tester. Two leads are connected to the disconnected and discharged capacitor terminals. An internal circuit loads the capacitor and indicates its condition by using a tone signal. The tone will indicate *good, shorted, open,* and *leaking.*

Troubleshooting an electrical circuit with capacitors will require some decisions on your part. Generally, with the exception of physical abuse or excessive voltage, capacitors are not the first candidates for testing. You may very successfully avoid testing individual capacitors and save time by that omission. On the other hand, if the capacitors have failed, they will adversely affect the circuit, and you may waste testing time until you come back to them for a satisfactory verification of their condition. With the exception of oscilloscope testing, when you are testing an individual capacitor, you need to isolate one of the capacitor's leads from the rest of the circuit, whether through an open switch or by disconnecting the lead itself.

Single-phase capacitor-start, or capacitor-run, motors that do not perform satisfactorily would require starting-winding testing that would include the capacitor. When a direct current power supply in an electrical panel fails, the diodes are the first place to look. However, the capacitors may also need verification.

TROUBLESHOOTING WITH PHASE METERS

A number of meters are available that indicate motor and phase rotation on three-phase wiring systems. These meters will indicate the direction of motor rotation after installation. They will also indicate the sequence of each of the three line phases. Strictly speaking, their value is in installation work more than in troubleshooting. Nonetheless, they are useful test instruments when the installation requirements merit them.

Reversing Rotation of Three-Phase Motors

One of the very practical advantages of three-phase motors is the ease of reversing their direction of rotation. A three-phase motor has three terminals. Reversing the connections of any two of these terminals will reverse the direction of the motor's rotation. If a three-phase motor shaft is turning clockwise when terminal 1 is connected to phase A, terminal 2 is connected to phase B, and terminal 3 is connected to phase C, it will turn counterclockwise if terminal 1 is connected to phase B, terminal 2 is connected to phase A, and terminal 3 remains connected to phase C.

It is obvious, however, that installing a three-phase motor requires careful attention to its wiring to avoid improper, and sometimes dangerous, motor reversal.

In many cases, a three-phase motor can be jogged (quickly started and stopped) to determine the direction of rotation before putting it into service. This presents no safety problems with an installation such as a reciprocating air compressor, since the motor can be briefly test-run in either direction. If the direction of rotation is incorrect, the unit can be shut down while two leads are reversed. There are applications, however, in which a motor cannot be reversed without risking danger to either the equipment or personnel. Elevators, hoists, augers, door openers, centrifugal pumps, and the like, may need to be correctly wired before any testing is done. If a hoist is wired in reverse, for example, the limit safety switches will not protect the system from overtravel.

Using a Phase-Rotation Indicator

There are two distinct test functions necessary for determining motor rotation. If you are installing a three-phase motor that must be ensured of correct rotation before starting, you will need to perform both tests.

1. The first test is phase rotation, which means that you are testing the incoming lines for their phase sequence. Properly generated three-phase power will sequence from phase AB to phase BC to phase CA. Properly installed feeder lines will correspond with this sequence.

2. The second test is motor rotation, which means that you are testing the motor for its direction of rotation after it is connected to the presently installed lines.

A number of manufacturers produce phase-rotation and motor-rotation meters. Figure 9–11 is an example of this type of test equipment.

Most meters use neon lamps to verify the condition of the lines and the phase sequence. Failure of any one of a first set of three lamps to glow indicates an open line. A second set of neon lamps verifies normal- or reversed-phase rotation.

The second test function is that of motor rotation. For this test, three test leads are connected to the disconnected motor terminals. With most meters, the meter test button is depressed while the motor shaft is turned in the desired direction of rotation. A glowing lamp will indicate the direction of rotation relative to phase connection.

For the actual motor installation, the motor terminals are wired to the corresponding lines according to the test results. That is, the known phase sequence of the incoming lines can be matched to the appropriate motor terminals when the motor test is correct. This will result in a known direction of motor rotation.

Once an electrician is familiar with the operation of any one of these meters, it should take less than a few minutes to test a motor connection during installation if the shaft is free to turn. Assuming that the necessary safety precautions have been taken, the actual energized leads at the motor coming from the magnetic starter could be connected to the phase-rotation portion of the meter. With the motor disconnected, the color-coded motor test leads would be connected to the motor terminals. The two tests would be performed, comparing the direction of rotation for each. If both tests indicated the same direction of rotation, the motor terminals and line wires would be connected according to the meter's color-coded test leads. If the test indicated a dissimilar direction of rotation, two motor terminals would be reversed when the terminals were connected to the line. Of course, the lines are de-energized after the testing—no installation work is done with energized lines.

The value of this type of testing before motor installation will greatly depend on the type and frequency of work you are doing as an electrician. Often, motor rotation can be determined by removing the link on the motor coupling. However, if you are working where startup testing is impossible or costly—or where reversed motors would be hazardous—the cost of phase-rotation meters is worthwhile. As motors—and conductors—get larger, the effort and cost of changing connections favors phase-rotation testing.

In keeping with this book's objective of reducing machine downtime and increasing plant productivity, phase-rotation test equipment may offer substantial time savings in specialized situations.

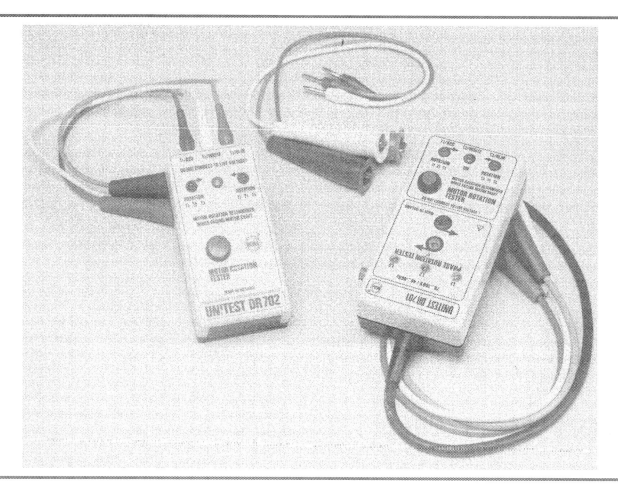

Figure 9-11 Greenlee Textron's 5774 Motor Rotation And Face Sequence Indicator contains the two test instruments show here. In addition to setting motor rotation before starting, the meters perform other tests such as identifying legs on three-face branch circuits. *Courtesy of Greenlee Textron.*

TROUBLESHOOTING WITH WIRE IDENTIFICATION INSTRUMENTS

There are a number of wire-sorting instruments available from various suppliers. A typical set is shown in Figure 9–12. Most instruments will have both a marker unit and an identifier unit. Typically, the *marker* unit has up to ten numbered leads and a common conductor. The *identifier* unit has one test lead and a common conductor. The *identifier* typically uses a LED number display from 0 to 9, which represents ten possibilities for wire identification.

Using a Wire-Sorting Instrument

Before using any wire sorter, it is important to verify that the circuits being tested are de-energized. The common leads of both the units are then grounded or connected to a known common wire. The actual testing is carried out by clipping the numbered leads of the *marker* to the unknown conductors in the circuit. When the other end of the circuit is touched with the *identifier* probe, the correct lead number will be displayed in the readout window.

There are many cases in which you might use a wire sorter. It can be used to verify wiring harnesses and thermocouple leads. It could be used it to identify pin locations in large, multiconductor (Amphenol™) connectors. In Chapter 11, you will learn how to identify unmarked wires with a long test lead and a meter. The wire sorter could make that job much simpler since you can identify a number of wires from a single hookup without the necessity of the long lead.

CHAPTER 9 Specialized Tests and Equipment **133**

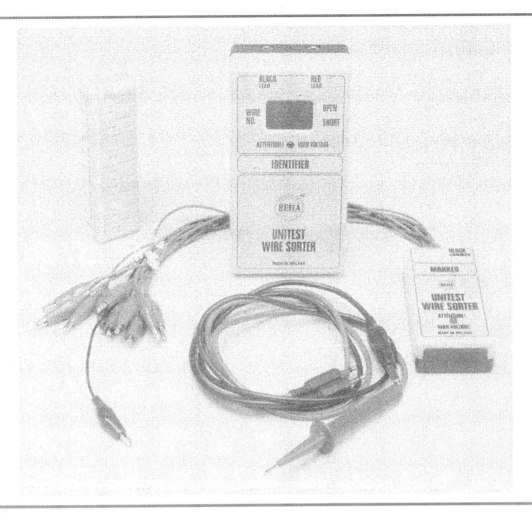

Figure 9-12 The Greenlee Textron 5775VS Wire Sorter is a typical wire identification instrument. These instruments identify wires from remote locations as well as performing other useful troubleshooting tests. *Courtesy of Greenlee Textron.*

Similar to the procedure of the last paragraph, you could determine the operation and wire identification of multiple-function switches. If you were testing a limit switch, you might connect the number 1 lead of the *marker* to the normally open side of the switch, and the number 2 lead to the normally closed side. The *identifier* clip would be placed on the common switch supply wire. You could verify that the switch was functioning correctly if the *identifier* changed from lead 1 to lead 2 as the switch was cycled back and forth. You could use the same procedure in more complex applications where you were verifying the operation of large numbers of conductors and switches. Of course, you would need to keep a notepad with number references since the ten numbers on the wire sorter (from 0 to 9) would not correspond with the actual wire numbers of the machine.

Be aware, however, that this type of instrument can only be used on de-energized circuits. Do not try to do on-line troubleshooting with this or any other type of specialized identification equipment.

Do you need a wire sorter? Needless to say, you can identify conductors just as accurately with a multimeter as you can with a wire sorter. However, you cannot do it nearly as quickly, and you will have more difficulty as the distance increases between your test points. Therefore, the question is really a matter of how frequently you would be using the wire sorter, and the total cost of the added downtime required for the slower procedure of using a multimeter. An added benefit with a wire sorter—as with any piece of specialized test equipment—is that you may discover other uses for it, such as the switch testing described above, which may justify its purchase price.

A DAY AT THE PLANT

Who needs a wire sorter? In answer, let me tell you an electrical horror story. I once saw a production machine that had been "moved" by a butcher—he had separated the entire control panel from the machine by hacksawing the umbilical leads. To correct that stroke of genius, you would need a wire sorter! If you were reconnecting the 200 plus wires—all the wires were conveniently color-coded "standard red"—you would undoubtedly start by identifying wires between the panel and the cut using the wire sorter. This would give you correctly numbered wires at the cut on the panel side. Then you would open covers and identify wire numbers on the machine by reading the wire identification labels. Using the wire sorter, you would then identify the wire number on the machine side of the cut and splice as you finished each identified pair. Any method would take time, but the wire sorter would allow you to work with groups of ten wires rather than with a single wire as you would be doing with a continuity tester.

Circuit Tracing

The larger and older your plant is, the more you will need to trace conductors. This will be particularly true in a multistoried plant, which centralizes electrical equipment in motor control centers and dedicated electrical rooms. A *dedicated electrical room* may be a crowded wall in the basement! Ideally, all wiring should be documented on prints and labeled on the physical equipment. In practice, this is not always the case. When old equipment is removed, control wiring will all too frequently be abandoned, leaving conduits with a mix of hot and disconnected conductors. Abandoning wiring is not a good practice, but you will encounter it someday.

On the other hand, you may simply be trying to find which disconnect in a poorly labeled motor control center de-energizes a conveyor circuit. Because of inadequate labeling and the need to avoid interrupting other equipment, you may still need to trace conductors. Circuit tracing in all of these less-than-desirable conditions broadly falls into one of two categories:

1. *Locating the source or termination point of known conductors.* In this instance, you are attempting to find the other end of a known wiring system. In all likelihood, you are trying to find the source. This may be true either of circuits in use or of disconnected wiring systems that are still in place.

A number of manufacturers provide current tracers. A transmitter is connected between the hot lead and a return ground path such as a neutral or conduit. Some units can be used on systems as high as 600 volts. It is assumed that the power can be de-energized while the transmitter is being connected. A hand-held receiver is then used to read a pulsing signal from the conductor. Most manufactures provide multiple sensitivity levels so that the receiver can read through conduit and walls, including reinforced concrete. On lower sensitivity levels, the receiver can identify specific circuit breakers and even individual wires in a bundle. Refer to Figure 9–13.

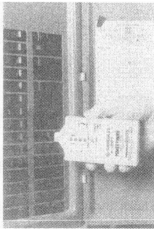

Figure 9-13 Two current Transfer in use. In both case, an energized conductor is being trance back to its appropriate circuit-breaker. These unit are manufactured by Greenlee Textron *Courtesy of Greenlee Textron.*

When the conductor is *not* energized, the better current tracers allow you to power the transmitter with a 9-volt battery from the termination end and trace the line toward the source. This also requires a ground path such as a neutral or conduit system. However, this is only done if the conductors are isolated from any possible voltage source.

Some current tracers can be used to locate wiring in a control panel. Figure 9-14 shows the connections you would use to locate a circuit for a particular control relay. The relay must be energized for this test. Some current tracers can be used for other applications, such as locating phase-to-phase shorts, locating underground water pipes, or tracing conduit.

Do not overlook the obvious in circuit tracing. If you are trying to find a specific motor controller that is one of twenty on a wall in the basement, there are several things you can do to simplify your search. The most obvious is that the controller will be sized for the motor. A 10-horsepower, 480-volt, three-phase motor will use a number 1 starter. You can at least eliminate the smaller and excessively large controllers. If the motor is in operation and you can safely measure current, the current draw on each of the three legs will match at both the motor and the controller. Probably the simplest procedure, however, is to identify the motor's switching by sound. With two-way radios, an assistant can tell you when the motor is cycling on and off as you listen for contactor noise. If you can disconnect the motor's line voltage with an in-sight lockout, have the assistant switch the controls on and off while you listen.

Similarly, a peak ammeter can be used to register cycling if you are conducting a momentary on/off test from a remote control station without an assistant. The ammeter can be clamped around the coil lead if the disconnect has been locked out. Do you remember the FLUKE 12B Continuity Capture function mentioned in Chapter 8? This meter could also be connected to the controller's coil to determine if the coil had been momentarily energized.

2. *Determining whether or not conductors are in use from mid-wire points.* In this instance, you

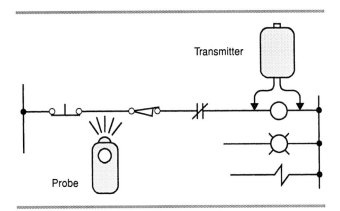

Figure 9-14 Some manufacturers' current trace can be used to locate wiring in a control panel. However, the circuit must be energized for the probe to locate the wire.

are attempting to identify conductors in junction boxes without knowing their source or termination point. Before you can remove old wiring, you must be able to determine if it is in use. A 1½-inch conduit may run between two or more locations in the plant passing through a number of pull boxes with interconnecting conduits. It may be filled with more than fifty number-14 control wires, gaining and losing some in each junction box. Without specific prints, how can you identify and trace individual conductors?

a. The most definitive evidence is current. If you have a clamp-on ammeter or a compact meter like the FLUKE T5-600, you know a wire is in use if it is drawing current.

b. Proximity testers are a second choice. Their indication is less reliable than current because they can give false readings from induction in wire bundles. In both cases, the *absence* of a current or electrical field is not definitive. It may indicate only that the circuit is temporarily off.

c. You could use a peak setting on a clamp-on ammeter to determine if a conductor was occasionally used. Periodically check the meter to see if any currents have registered. Again, the same caution must be made; absence of a current does not mean that the conductor is disconnected from a power source.

d. At some point you must do a physical trace. In most instances, this is a quick procedure that will take you to an identifiable source or termination point. As the complexity increases, you will need to be more careful in identifying the color and size of the wires, the number of wires in a given conduit, the run and size of the conduit, and so forth.

Safety must be a constant concern when tracing conductors. If the conductors are in service, you must be certain that they are de-energized and locked out before connecting tracers or performing any work. This may be done on the equipment's disconnect before tracing the conductor back to the unknown source. You must also ensure that the conductors will not be unexpectedly energized. When you are removing old wiring, you must be certain that it is de-energized and not merely temporarily off. Always carry a proximity meter to test conductors before cutting.

Get in the habit of cutting one wire at a time; inadvertently cutting a hot wire in a bundle can cause a short circuit or energize other wiring. Wear appropriate electrician's gloves when you work. If the conductors are not fully identified, after cutting a wire, place a small wire nut that will grip the wire jacket firmly on the source end of the cut wire. Pay particular attention to energized conductors in the work area. When you are tracing or removing old wiring, you will frequently be working close to other circuits.

INNOVATIVE ELECTRICAL TROUBLESHOOTING

The real purpose of this chapter is to force you to think creatively when you are using electrical test equipment. When you purchase your new meter, you will read the instruction manual thoroughly. That instruction manual will demonstrate all the tests your new meter can do in conventional troubleshooting settings. It can perform test functions such as determining voltage values, detecting continuity, sorting unidentified wires, or displaying an insulation value.

You will be testing those same values yourself. But are you using those values innovatively to reduce your troubleshooting time? Finding a low-insulation value is a normal test for a megger. But finding that low-insulation value in a shorted limit switch is innovative when it reduces the troubleshooting time to less than ten minutes. Using a "peak" clamp-on ammeter to determine the highest amperage for a running motor is routine. To use it to verify that an expensive circuit-breaker is truly at fault to avoid unnecessary expenses is innovative. Current testing in an electrical panel is done every day. Using a clamp-on ammeter to determine whether a hydraulic solenoid is properly closing in order to avoid disassembling a hydraulic valve is the work of a maintenance artist.

Learn to evaluate your troubleshooting problems. Carefully determine which electrical values (voltage, continuity, high or low resistance, current, frequency, and the like) will best identify the problem you are searching for. Then use both your test equipment and the machine itself to identify those values.

But work safely. Innovation does not mean you can use your test equipment carelessly or ignore accepted safety procedures. *Never* compromise safety for either yourself or the machinery you are troubleshooting.

TROUBLESHOOTING WITH A NEON LAMP

After discussing expensive solid-state equipment that can analyze every fault in the electrical circuits you are testing, it is time to come back to the basics.

If you are on-line troubleshooting a faulty circuit, you are simply trying to find one thing—an "open" in the circuit. In Chapter 8, you were shown how to use a multimeter to locate the problem.

If a multimeter will give you your troubleshooting information by indicating the presence of a voltage, why can you not do the same thing with nothing more than a $3 neon glow lamp? The glow lamp is simply a resistor-protected neon lamp in a molded plastic case. It has two 6–inch leads that are touched on the bare conductors. It will glow if it is across 60 volts or more.

We are not suggesting that all which has been said in praise of electronic test meters is oversell. Any electrician would dread having nothing but a neon lamp available for his or her own troubleshooting work. At the same time, as important as good instruments are, the skill of the troubleshooting electrician is still the most valuable asset for effective diagnostic work.

Could you troubleshoot with only a neon test lamp? Yes! You could use a neon test lamp and exactly duplicate the first three troubleshooting steps given in Chapter 8. A knowledgeable electrician might do more with only the test lamp than a poorly prepared electrician could do with a box full of test instruments. What, then, could a *knowledgeable* electrician do with a box full of test instruments?

CHAPTER REVIEW

Troubleshooting speed can often be increased through a careful selection of specialized testing equipment in conjunction with a thoughtful and innovative approach to its use. Even though this chapter merely touched on some possible troubleshooting alternatives, their suggestion should supply a basis for the development of innovative testing techniques in your specific field of application.

A clamp-on ammeter can often be used to determine if the electrical power component or output device at the end of the circuit is functioning normally. The presence of a voltage to the component's terminals is often inconclusive, whereas a current value measured with an ammeter can give a more accurate indication of the condition of the power component. When the ammeter is capable of reading very small current values, its applications for troubleshooting electrical controls in a panel or on the machine are greatly enhanced. Examples of diagnostic troubleshooting were given where the current values of improperly functioning relays or solenoids were compared with normal values. Examples of specialized ammeter functions such as "peak" value holding were given with an explanation of their value in on-line troubleshooting work.

Insulation testing equipment such as megohmmeters are useful for troubleshooting work. You were encouraged to take phase-to-ground readings from the most accessible site on the machinery. In most cases, this would be from the motor contactor or a fused disconnect. By taking your first readings from these readily accessible areas, a great deal of time can be saved by avoiding the unnecessary location of testing terminals on the motor itself. The visual and meter inspection of the actual motor should take place only after there is reasonable certainty that the motor is at fault.

A general discussion of the cause of motor burnout identified four major causes, all dealing with insulation failure. They are:

1. Heat-related insulation breakdown
2. Moisture-related insulation breakdown
3. Mechanical-related insulation breakdown
4. Oil- or chemical-related insulation breakdown

Insulation failure can be determined with the use of the megohmmeter. This testing can often be done from points outside the motor such as the motor contactor. Because of the high inherent test voltage output of the megohmmeter (500 to 1,000 volts), this test instrument must never be used on capacitors or solid-state circuits.

Capacitors are constructed with two conductive plates separated by an insulating dielectric. Capacitor failure is generally in the form of a breakdown in the dielectric or a short between the plates. Capacitors must be handled carefully since they can hold a charge that can damage test equipment or be hazardous to personnel if the terminals are not first shorted. Special capacitor testers are available that will indicate the various fault conditions of a capacitor.

Other specialized equipment is available, which can be an asset in troubleshooting work. Phase-rotation meters can be used to identify leads and establish motor rotation. A wire identification meter can be used for various troubleshooting problems beyond mere wire numbering. This includes identification of control wires in other parts of the circuit, identifying thermocouple leads and wire harness terminals, and similar types of identification work.

An underlying purpose of this chapter is to encourage you to creatively use voltage, amperage, resistance, and other test values to enhance your troubleshooting effectiveness. However, innovation must never compromise safety.

The chapter closed by taking you back to the simplest of testing tools—the neon lamp. The purpose was to remind you that the measure of the electrician is not in the quantity of specialized tools he or she carries. Rather, the true measure is in the electrician's ability to use knowledge in conjunction with the necessary tools to come to quick and effective troubleshooting solutions.

THINKING THROUGH THE TEXT

1. Why is the presence of a voltage on the wire leading to an electrical component, or output device, inconclusive evidence that the component itself is functioning?

2. What three test values can you work with in on-line troubleshooting? What does each value indicate?

3. What type of meter would be used to indicate that a solenoid or relay was not moving to a fully closed position? How would the reading vary from a normal reading?

4. Irrespective of the initial outside condition that causes a motor problem, what is generally the immediate cause of motor failure?

5. When testing a motor for insulation failure, two tests are conducted with an ohmmeter or a megohmmeter. These are phase-to-phase and phase-to-ground tests. Where are the meter's test leads placed for these two tests?

6. What are the four broad causes of insulation failure?

7. Two testing conditions are named in which a megohmmeter must never be used. What are they? What is the reason a megohmmeter cannot be used for these tests?

8. What are the four most frequent uses of capacitors in industrial settings? What caution is given before handling or testing capacitors?

9. What are the two primary causes of capacitor failure? Briefly explain each.

10. How do you reverse the direction of rotation of a three-phase motor?

11. What does the author identify as the most valuable asset for effective troubleshooting?

CHAPTER

10

Troubleshooting with Industrial Oscilloscopes

OBJECTIVES

After completing this chapter, you should be able to:

- Understand the basic functions and use of an industrial-rated hand-held oscilloscope.
- Describe the difference between an *instantaneous waveform* and a *trend plot.*
- Use an industrial hand-held oscilloscope to monitor both the electrical and mechanical characteristics of production equipment over time.
- Develop testing and recording procedures for an industrial hand-held oscilloscope that you could use in your own plant maintenance program.
- Identify the safety practices required for the use of a hand-held oscilloscope in an industrial setting.

INDUSTRIAL-RATED OSCILLOSCOPES

Several manufacturers produce hand-held oscilloscopes for use in industrial troubleshooting. In addition to their small size and battery operation, these meters are characterized by their high-voltage withstand ratings and isolation from the harmful effects of electrically noisy environments.

All testing for this chapter was done with a FLUKE 123 ScopeMeter®, which is shown in Figure 10–1. This meter has a 600–volt working limit and is designed for use in CAT III industrial environments. This particular ScopeMeter model was selected because of its mid-range price with the purpose to demonstrate the usefulness of a hand-held oscilloscope within the equipment budget of an industrial plant. Other more expensive ScopeMeter models have additional functions.

Testing Orientation

Industrial electricians often shy away from technically demanding testing equipment such as oscilloscopes. However, knowing either the **waveforms** or the voltage and amperage trend plot produced by plant equipment can often reduce downtime. The ScopeMeter used in producing this chapter was used as an *electrician's* test instrument. Though the ScopeMeter is capable of many measurements within electronics (such as decibels, capacitance,

140 CHAPTER 10 Troubleshooting with Industrial Oscilloscopes

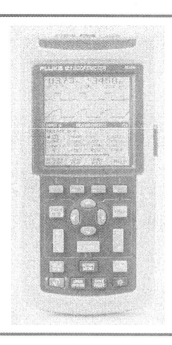

Figure 10-1   The FLUKE 123 ScopeMeter® described in this chapter. *Courtesy of FLUKE Corporation.*

triggering on video signals, and the like), those functions were not emphasized for the tests described here. Rather, as an electrician, you need to appreciate the value of knowing the waveforms produced by electrical machinery during operation.

Variable speed drives have become commonplace in the industrial setting. An oscilloscope is a necessary test instrument for these units. This chapter includes an example of a frequent test that a plant electrician may routinely use to verify drives in operation.

Conventional bench oscilloscopes are too delicate and cumbersome for production plant use. The ScopeMeter, however, is made for this harsh environment. It can be directly connected to 480–volt circuits. The ScopeMeter can also tolerate the extreme electrical noise in the industrial plant setting. We have monitored continuous 700–ampere welding operations on a pipe mill with a plasma cutting torch in operation. A conventional oscilloscope was used earlier on the same pipe mill and lost its power supply when the plasma arc machine fired!

All alternating current voltages and currents displayed on the ScopeMeter are true RMS values.

The majority of our ScopeMeter testing will involve three test modes:

1. The first test mode displays the *instantaneous waveform* of voltage and/or amperage of equipment in operation. *Waveform* simply means the graphic display of the voltage (or current) relative to a reference point. For an alternating current waveform, this will generally be a sine wave beginning at a zero reference, rising to a peak, and again falling to the zero reference point. This trace is repeated for both a positive and a negative voltage value for each cycle. This is the waveform shown in the AUTO mode on the ScopeMeter. *Instantaneous* means the waveform that is actually occurring at the time the reading is taken.

2. The second test mode displays a *trend-plot-over-time* of voltage and/or amperage of equipment in operation. The *trend plot* function plots the voltage (or current) readings as a function of time. An unvarying 120–volt alternating current voltage value will appear as a straight line at 120–volt increments above the zero reference line. *Over time* means that the resulting trace will be displayed relative to the time of occurrence. This is the trace which is shown using the TrendPlot™ function on the ScopeMeter.

3. The third test mode monitors the ***power quality*** of the plant's electrical system. Generally, this is a specialized use of the first test that evaluates the waveform for evidence of harmonics and other power distortion.

In addition, this information will be used in one of two applications:

1. The resulting measurements will be used for *immediate diagnostic purposes* when troubleshooting faulty equipment.

2. The information will be logged for specific production equipment so that *performance-over-time* may be monitored. In this second application, a trend plot made at a prior time will be compared with a contemporary trend plot for the purpose of evaluating machine performance.

Because the waveforms or traces from the following examples may be either used for immediate diagnostic purposes or logged for future evaluation of production equipment, the specific application will not be identified for each example. The connections and meter settings for the following tests are not described.

Reading an Oscilloscope Screen

Because many of the figures in this chapter are taken directly from an oscilloscope, an explanation follows of the information displayed on the screen. You can selectively show either one or two traces on a two-channel scope. The traces are generated simultaneously so that you can compare the action of a second value (amperage, for example) with that of the first value (voltage, for example). Figure 10–2 shows a screen displaying both a voltage trace (top, channel A) and a current trace (bottom, channel B). Numeric values of the measurement are given on the top of the screen. Channel A is showing a peak value of 119.3 RMS alternating current volts. The smaller numbers underneath are showing that there is a +000.0–volt direct current component. As electrical noise increases, there are often foreign direct current components in an alternating current voltage (or vice versa). Channel B is showing a peak amperage of 1.79 RMS alternating current amperes and a direct current component of 1.79 amperes. Because the oscilloscope gives numeric values, the scope can be used in the same way as a DMM for reading voltage and current values.

The division values are identified at the bottom of the screen. The A channel is displaying 100 V/d, which means that each vertical division represents 100 volts. Since the peak voltage for 60 hertz alternating current is the given voltage times 1.414, the peak voltage for 119.3 is 168.69 volts (119.3 × 1.414 = 168.69). On the sixth division line from the bottom, on the left-hand side of the screen, channel A displays a heavy marker on the zero line. You can see that the crest of the voltage trace covers approximately 70 percent of the second division space, or 170 volts from the marker on the zero line. Equally, the negative (bottom) portion of the trace falls at approximately 170 negative volts from the zero line.

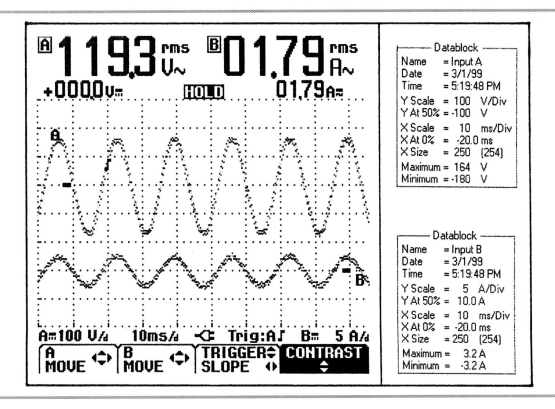

Figure 10-2 The FLUKE 123 ScopeMeter screen. Numeric values are given on the top of the screen as averages; the trace is displayed in the center of the screen; division values are given at the bottom of the scree. Sidebars give maximum/minimum data rather than averages.

The bottom of the screen also identifies the horizontal time divisions. There are ten milliseconds per division: 10ms/d. The time required for one complete cycle, which includes one positive and one negative crest, is therefore 1/60 of a second. The screen is displaying 1/10 of a second.

The B channel is identified as displaying 5 amperes per division: 5A/d. When you look at the bottom trace, you can see that the crest rises to 1.79 as measured from the heavy marker on the right-hand side of the screen on the second line from the bottom. Similarly, the current trace alternately shows a negative current of 1.79 amperes.

In use, you can independently adjust the vertical and the horizontal scale divisions to best accommodate the trace you are viewing. You can also move the trace location. In this example, we separated the two traces, placing one on top of the other. You could also use a common zero line and superimpose the two traces.

Now that you understand the screen's division marks, you can see the significance of the trace(s) displayed on the screen. When you use an oscilloscope with numeric displays, you will probably not use the trace for computing voltage values unless you are identifying peak voltages and the like. However, the trace will give you valuable information concerning the electrical characteristics of the power you are testing. The two traces in Figure 10–2 are from a clean 120–volt power source with a purely resistive (non-inductive) load. However, in the examples that follow, you will see how the oscilloscope can display many types of power distortions and noise that cannot be determined by reading the voltage or amperage values from a DMM's digital display.

A Note about Safety

Throughout this chapter, electrical connections made with the ScopeMeter are described. These tests are made on live circuits. In all cases, *de-energize the circuits before making meter connections*. Make certain that all connections are secure before reenergizing the circuit. Hang or support the ScopeMeter so that the meter is not held in the hands during live voltage testing. Touching the on/off or hold buttons will be the only contact with the meter and leads during on-line testing. These precautions apply to all line voltage testing in this chapter, though they will not be repeated for each procedure.

For proper protection of the meter, make certain that the measured voltage will not exceed the design limits of the meter itself.

WAVEFORM TESTING: GENERAL DIAGNOSTICS

Waveform Example #1: Local Power Distortion

As a simple first example, the waveform that comes from a small inductive load during switching will be evaluated. A small alternating current power relay, which draws approximately 0.86 ampere at **inrush** and 0.17 ampere when **sealed**, was momentarily switched on and off for this example. A push button switch was momentarily touched in order to capture both the sealing and release points on the same graph. Even though this produced a cycle frequency that was considerably higher than normal, the waveforms are representative of normal operation. As you can see, the waveform is symmetrical between the switching points. In an electrical panel with only electromechanical relays, timers, and motor controls, we would not be concerned with these switching transients. However, when sensitive electronic equipment is incorporated into panels, care must be taken to properly isolate them from the transients produced by relays and motor controllers. The simple test across the coil terminals shown in Figure 10–3 indicates that this small coil is capable of producing transients in excess of 800 volts. Typically, the high voltage values are random. Often, the peak value was no more than 300 volts in the tests conducted to produce this example. In some instances, there was no appreciable voltage rise at all. The voltage intensity will depend on the sine wave position at the actual time the contact is opened. Notice also that the largest spike is produced when the contacts are *opened*. Think back to the old automotive distributor point ignition system. The high voltage fed to the spark plugs was produced when the ignition points *opened* and the spark coil field collapsed.

These short duration bursts of 800 volts shown in Figure 10–3 can destroy the sensitive solid-state

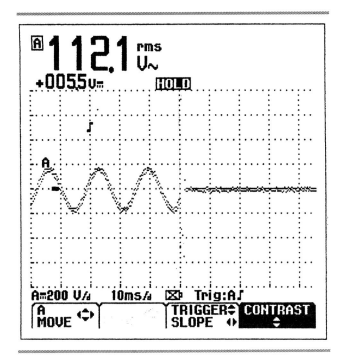

Figure 10-3 Power distortion produced by a small relay coil during switching

circuits in timers, counters, proximity devices, and the like. Devices such as **resistor/capacitor (RC) snubbers** or **metal oxide varistors (MOVs)** are usually installed across alternating current relay and motor controller coils to dampen transients. The generic term for these devices is a **surge suppressor**. The device is mounted in parallel with the coil with one lead of the device directly mounted to each coil terminal. A diode provides the same protection on direct current relays.

For this test, the ScopeMeter Trigger (under Scope options) was activated. This function holds the waveform on the screen when abnormal events such as transients occur. This same test could be used in the field. If, for instance, an electronic device was being installed in a control panel, this test would determine the suitability of the control power leads chosen as the eventual power source. If necessary, individual relays and motor controllers could be cycled to identify specific sources of transients. The test could be conducted both before and after appropriate RC snubbers were installed. All testing is completed, however, before power is connected to the electronic device itself.

Waveform Example #2: Welding Machines

Many types of electrical equipment in a plant setting use a waveform that varies from the sinusoidal 60–hertz alternating current wave supplied to the plant. This is true of all equipment using a direct current output from simple battery chargers to direct current variable speed drives or direct current welding equipment. It is equally true of alternating current (inverter) variable speed drives and other alternating current equipment. The resulting waveforms can often be used to give important diagnostic information regarding the condition or performance of specific equipment. Many of the following examples come from welding machines. The units used for these tests were direct current output machines ranging from 400 to 1,000 amperes. They were generally connected to a 480–volt alternating current power source. These machines all used some form of rectification to convert the alternating current input power to usable direct current welding power. Newer equipment usually uses **silicon-controlled rectifiers (SCRs)** rather than diodes because the SCR can be controlled.

The waveforms on both the input (alternating current) side and the output (direct current) side of a welding machine offer valuable diagnostic information. When evaluating output, it is desirable to maintain a constant voltage and current. As you will see later in Figure 10–7, neither voltage nor current are constant under actual welding conditions. To maintain these constant values for test purposes, a large resistance bank is connected to the output leads of the welding machine. Figure 10–4 shows an air-cooled load bank manufactured by the Miller Welding Equipment Company that is used for this purpose. When all rectification is operating normally, the waveform pattern should be smooth and symmetrical as shown in Figure 10–5. In the following examples, trace A is always voltage and trace B (when used) is amperage unless designated otherwise.

However, when rectification is faulty, the wave pattern will be nonsymmetrical between phases. The direct current output on a three-phase welding machine displays the rectification of the individual phases. Figure 10–6A shows the alternating current input trace. Though the 480–volt trace is symmetrical,

144 CHAPTER 10 Troubleshooting with Industrial Oscilloscopes

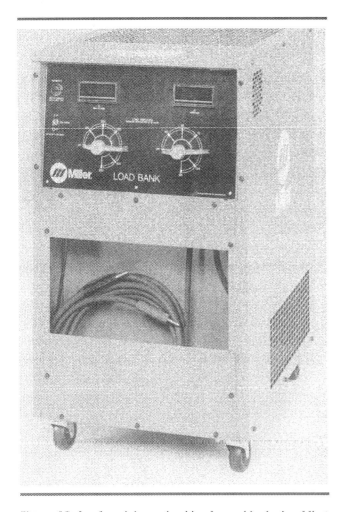

Figure 10-4 A resistance load bank used to test welding machines under full load. *Courtesy of Miller Welding Equipment Company.*

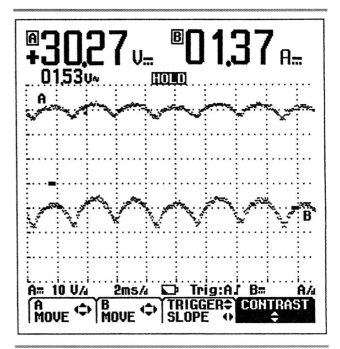

Figure 10-5 The output waveform of a normally operating DC welder. Both the voltage waveform (trace A on top) and the amperage waveform (trace B on the bottom) are uniform.

the *current* trace is not. When you look at the output trace in Figure 10-6B, you see that both the output voltage and the output current show a nonsymmetrical pattern. From these patterns, you could conclude that something in either the control board or the actual rectification itself may be operating improperly.

These traces are used as examples because they are simple to visualize. Modern welding machines do not always show symmetrical waveforms. A welding machine may fire one SCR at a fixed rate and control two SCRs at variable rates. In practice, the best diagnostic information will come from watching a particular machine over time to determine *changes* in the waveform. On the other hand, when an SCR has failed completely, characteristic flat lines (from an open SCR) or spikes (from a shorted SCR) will be obvious.

For the alternating current input voltage trace in Figure 10-6A, all connections were made in the de-energized fused disconnect. The clamp-on current probe was placed around one phase conductor and the two voltage leads were attached to the fuse holder spring clips. Proper safety precautions were taken in re-energizing the circuit with the disconnect door ajar. The direct current traces were recorded by connecting the ScopeMeter lead A to the positive welding machine terminal post, and the COMMON lead to the negative post. The clamp-on current probe was placed around an electrode lead. Both the voltage and the current probe must be properly oriented or the direct current waveforms will be inverted. On the 35-volt output side of the welding machine, improper connections can be corrected with the machine running, or by reversing the polarity of the displayed waveform on the meter itself. For all traces in Figures 10-5 and 10-6, the ScopeMeter was used in the AUTO mode with input A set on VAC and input B set on Amp.

An industrially hardened oscilloscope such as the ScopeMeter has the obvious advantage of usefulness in hostile electrical environments. Consequently, evaluations of the actual welding process can be

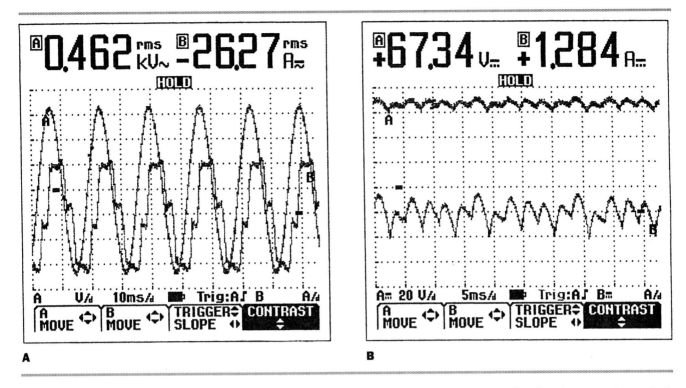

Figure 10-6 (A) The AC input to a defective welding machine. The voltage trace is symmetrical but the current trace indicates imbalance between the phases. (B) The DC output from the same machine. Notice that the current trace B is not equally loaded. Rectification is faulty.

safely conducted. In Figure 10-5 and Figure 10-6B, the output of welders was evaluated using a fixed resistance. Similar waveforms can be obtained with welders in actual production. Figure 10-7A shows a welder with acceptable rectification in operation. Figure 10-7B shows an identical make and model of machine in operation. Notice the difference in the two waveforms. The voltage trace of Figure 10-7B shows very short duration spikes that correspond with disruptions in the current plot. The actual welds were being monitored in this example. The two tests were done by the same experienced welding operator. Both tests were done with the same welding leads and wire feeder, and were conducted within 15 minutes of each other on the same base metal. Consequently, as many independent variables as possible were eliminated. Visually, both welds from Figures 10-7A and 10-7B were acceptable. Though X-rays of the comparison welds were not taken, the welding machine in Figure 10-7B is suspect; it merits close scrutiny. In the future, time and experience will allow the electrician and quality assurance team to define those waveforms that accompany field weld tests that fail to meet performance standards.

Waveform Example #3: Variable Speed Drive Testing

One of the practical applications for an oscilloscope is the verification of the output of a variable speed drive. Alternating current variable speed drives, which are typically known as **variable frequency drives** (**VFDs**) (they are also called **inverters**), are able to control motor speed by varying the output frequency. The internal electronics of the drive itself rectify the incoming alternating current power to direct current. The direct current is then fed back into Silicon Control Rectifier (SCR) circuits, which develop a variable alternating current waveform. Notice how the waveform in Figure 10-8A is actually built from a large number of direct current segments. However, because the new waveform is independent of the incoming alternating current frequency, the electronic circuit is able to build an alternating current waveform of any required frequency. Thus, for example, a frequency drive alternating current motor that is rated as a 1,750 rpm motor at 60 hertz will run at approximately 875 rpm when the VFD has an output of 30 hertz. The same motor will run at approximately 2,200 rpm if the VFD output is 75 hertz.

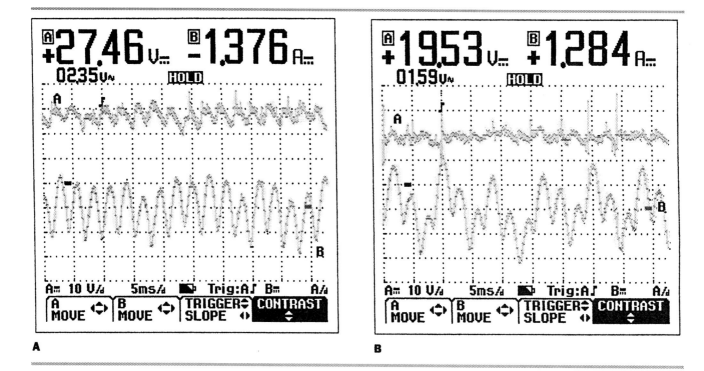

Figure 10-7 (A) The DC output of a welding machine in operation with acceptable rectification. Both the voltage A and the amperage B the second graph show the DC output from an identical machine in operation with faulty rectification. The voltage trace A shows a regular spike and the correct trace B indicates that each phase is carrying different amperage.

As simply a matter of interest, the VFD has an extremely practical application in motor installations that require frequent starting and stopping. In a typical alternating current motor application, the motor will draw some 600 percent of its full-load running current at start because the motor field has a constant 60 hertz rotation, whereas the armature is accelerating from zero to full speed. This high initial slip ultimately results in high winding temperature. For this reason, most alternating current motors are rated at a maximum number of starts permissible over a given time. On the other hand, most VFDs have provision for ramping the speed. This means that at start, the VFD progressively goes from 0 hertz, to 1 hertz, to 2 hertz, to 3 hertz, and so on, until the drive finally reaches the set operating speed. Because the motor is within its normal percentage of slip at each frequency setting, undue heat from excessive current is avoided. Thus, a VFD can be installed where a motor must start and stop frequently. For example, a VFD drive could be used on a carriage used to load product into a manufacturing process. The VFD could be programmed to start and accelerate the carriage to full speed after loading, then decelerate and stop the carriage at the unloading station. It would then reverse through an accelerate/decelerate program and bring the carriage back to the loading position. This could be done as many times an hour without problems for the motor as the physical size of the carriage and load would permit. This installation would be simpler, requiring less maintenance than older style electric drives, which required clutches and brakes on a continuously running, constant rpm motor. For similar reasons, VFDs have largely replaced the older-style variable speed systems (called Vari-drives), which used a belt between two spring-loaded variable diameter pulleys. Belt maintenance was quite high on these older mechanical units.

There are two types of oscilloscope testing on VFDs. Diagnostic and repair work on the drive itself would obviously require an oscilloscope along with a high degree of technical training. In this text, we will not attempt to show any of the testing procedures used in the repair of a drive. This is the work of a drive technician. On the other hand, the second type of testing—verification of drive output—is one that every plant electrician could use if VFDs are in use.

been in operation for a number of years without problems. Then, for a period of several months, the fuses would infrequently blow when the motors restarted after the defrost cycle. Finally, in about a week's time, the fuses started blowing each time the motors restarted. However, if the fuses were changed, the fans would operate without problems until the next defrost cycle. The troubleshooting procedure was as simple as comparing the waveform of each drive. After the fans had been restarted, a clamp-on probe was placed around each of the three VFD's outputs. (A current measurement was used because it required no direct contact with the 480–volt output.) The ScopeMeter waveform function was used. Two drives gave the output of Figure 10–8A. One drive gave the output of Figure 10–8B. Clearly, the VFD drive of Figure 10–8B was defective. By switching the defective drive off, the system was used without further loss of fuses until the drive was eventually replaced.

TREND PLOT TESTING: EVALUATING EQUIPMENT OVER TIME

Trend Plot Example #1: Motor Starting Current

Motor starting current can be easily monitored with the ScopeMeter trend plot function. Place the current clamp around one motor lead, set the meter to read alternating current amperes, and start the motor with its normal load. Figure 10–9 was obtained on a motor with a light starting load. A heavy starting load would show slower acceleration on the high current peak. This figure also shows that the motor was turned off at the conclusion of the test.

When using the ScopeMeter with FLUKE's software program, the display may be enhanced on a computer screen. By using this option, we can determine that the peak current was 134 amperes by reading either the scale divisions themselves or the sidebar information.

Trend Plot Example #2: Connection Resistance

All electrical installations have at least one branch circuit feeding the equipment. The branch circuit—

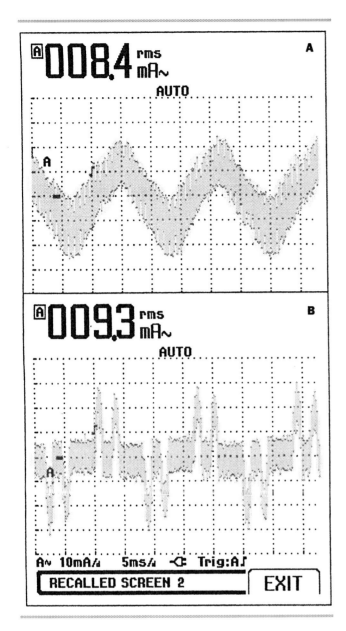

Figure 10–8 (A) The normal output of a variable speed drive (inverter) operating on 480 volts (top). (B) The output of a defective variable speed drive (bottom).

Figure 10–8 comes from an actual test in a dairy processing plant. In a roof-mounted building above a cold storage area, this plant used three VFDs to control three 5–horsepower fan motors on an ammonia compressor evaporator. When the system was installed, a single set of fuses supplied three identical VFDs, each rated at 5 horsepower. Individually fused drives would have been a better installation choice. Once each 24 hours, the system automatically switches to a 30–minute defrost cycle during which time the fans are stopped. The system had

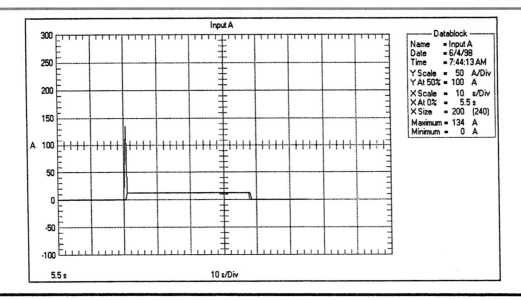

Figure 10-9 The plot of a lightly loaded motor during starting. By using Fluke's ScopeMeter software with a computer, the graph can be enhanced with absolute values displayed numerically in the sidebar.

as well as individual circuits within the machine itself—will pass through a number of low-resistance (conductive) connections. The typical fused disconnect has two lug connections, two knife switch connections (either a hinge and knife or two "points"), and two fuse clip connections *for each phase*. Consequently, in most three-phase fused disconnects alone, there are eighteen connection points that could deteriorate and cause erratic voltage drops to the power loads. Multiply that by the large number of connections in a manufacturing plant installation, and the potential for poor conductor connections is readily apparent.

However, the problem is further complicated because high-resistance connection failures initially occur only at full operating current. It is only after complete failure that visual inspections will reveal heat-damaged lugs and insulation. In today's manufacturing plant, the variable speed motor drives and PLC-controlled equipment will not tolerate voltage and current fluctuations from high-resistance connections.

The ScopeMeter can readily be used to determine voltage fluctuations from poor connections.

As a result of a one-time painting process, the three-phase runway contact conductors of an outdoor 5-ton bridge crane were fouled with epoxy paint. When the painting process was completed, the conductors required cleaning so that the crane could be put back into service. The outdoor portion of the crane travel was over 150 feet long. The crane operated at 480 volts. Because a ScopeMeter was available, the contact between the collector wheels and the runway conductors was verified after cleaning.

With the main runway disconnect locked out, the ScopeMeter was attached to the load side of the opened disconnect in the crane cab. This location placed the ScopeMeter after the trolleys riding on the overhead conductors. Test probe A was connected to the phase A load conductor in the fused disconnect. Test probe B was connected to the phase B load conductor. The COMMON (ground) test probe was connected to phase C.

The testing was done from inside the crane cab itself. After the crane bridge travel was started, the ScopeMeter was turned on in the trend plot mode. The bridge was held in constant motion until the HOLD button was pressed to stop the plot. The result was the trend plot shown in Figure 10–10.

Before looking at the resulting test pattern itself, you need to know how the three-phase conductors were monitored with two ScopeMeter inputs. With optimal contact between the trolley and conductor of phase A (represented by the top line on the graph) and the trolley and conductor of phase C, there would be a continuous horizontal line at the

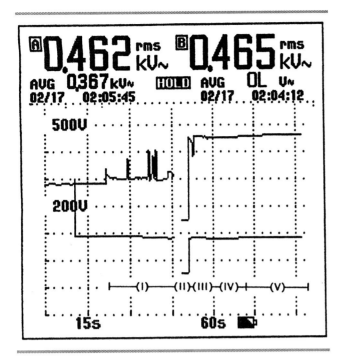

Figure 10-10 The plot of three crane runway contactors in motion. Phase A is represented by the A plot. Phase B is represented by the B plot. Phase C is represented by symmetrical movement in both A and B plots.

480-volt level because of the 480-volt potential between phase A (the A input lead) and phase C (the common input lead). Therefore, any asymmetrical alteration in the top line indicates the voltage variation between the trolley and conductor for phase A. Similarly, with optimal contact between the trolley and conductor for phase B and the trolley and conductor for phase C, the test pattern will show a continuous horizontal line for the second trace because there will be a 480-volt potential between the trolley and conductor of phase B and the trolley riding on the conductor for phase C. However, if phase C breaks contact, the potential difference between both C and A, and C and B, will alter symmetrically. Therefore, both the A and B phase contacts may be read directly; the C phase contact is interpolated from the simultaneous variation in phases A and B.

From the test pattern of Figure 10-10, all three runway contactors were verified for approximately 75 feet of crane travel over the paint-affected area. Figure 10-10 indicates that there was poor contact in section (I) of the phase A conductor. It also shows that the same portion for both phase B and C was operating normally. Section (II) shows that phase C momentarily lost all contact, and the crane was coasting because both phase A and B changed symmetrically. Section (III) shows that phase B and C were conducting, but phase A was making poor contact. At section (IV), both phase A and B were making satisfactory contact, but phase C was encountering slight resistance. In section (V), all phases were making good contact. The rise shown in the plot for phase A after section (I) was the result of the meter autoranging to a different scale after initial start-up.

This same test procedure could be duplicated on any three-phase machine encountering power fluctuation problems. It could also be done on single-phase equipment by using the ground as reference to two hot legs. Say, for example, that a small batch plant and conveyor system was occasionally running erratically, with the problem affecting the whole system and not just a single motor. However, after a brief interval, the problem would clear and it would run satisfactorily. This test could be run on the branch (or feeder) circuit to the batching area. There is nothing that would prevent this test being done on a single motor that was giving intermittent service, either.

Remember the goal of on-line troubleshooting. You do not want to unnecessarily shut equipment down to do your testing. Consequently, over a break when the equipment is normally idle, you would lock out the equipment and attach your ScopeMeter at a point where you know the wiring is common to the area having the problem. When the machinery restarts after break, you touch the "on" button of your meter and close the disconnect door (or secure the area if the disconnect must be left ajar). You then leave instructions that the operator is to notify you immediately when the equipment again runs erratically.

Twenty minutes later, the call comes. After you reach the batch plant, you watch the trend plot. As expected, you see a ragged plot on one phase (or both phases symmetrically). You now know that you have a poor connection somewhere in the system. You may be able to find the faulty connection by visually inspecting every connection point up-line from the batch plant. You will be aware of anything that is hot or smells like burned insulation. You will also be looking at any conduit or panels, which may have evidence of physical damage. But you are not *touching* or *pulling* on wires until the circuits you will be working on are *locked out!* If you cannot find the problem in your initial inspection, you could duplicate your original test by moving up-line toward the service panel. If at some point the equipment is running poorly, but it does not show on the trend plot, you know the poor connection is now downline of your test.

Equipment with a faulty connection will not operate long before there are more serious consequences. But early diagnosis will often define the problem you are looking for before complete failure and shutdown of the equipment results. Secondly, early diagnosis can often greatly reduce downtime by minimizing damage. Tightening a nut in a disconnect may be all that is needed if it is found early. Nonetheless, if damage has been done, replacing a disconnect still takes less time than pulling new conductors because the original wire became unusable with charred insulation next to the lug.

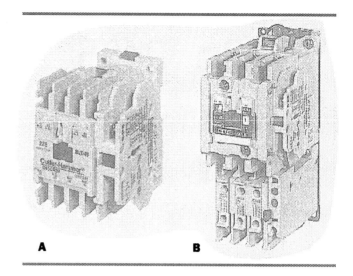

Figure 10-11 Industrial relays and controllers fall into two broad categories: (A) Low-current-carrying control relays; an example is shown on the left. (B) High-current-carrying relays and controllers; a motor controller is shown on the right. Courtesy of Cutler-Hammer.

Trend Plot Example #3: Motor Contactor and Relay Points

Mechanical contact points generally fall into one of two categories. A first category consists of relays used for control circuits. Generally, these relays are rated at 10 amperes or less. There are many styles, though the plug-in relays (often called "ice cube" relays) and the multipole industrial control relays are probably the most frequently encountered. A multipole relay is shown in Figure 10–11A.

Motor controllers (or power relays) shown in Figure 10–11B represent the second category of mechanical contact points. The contacts in these units are rated from 3 amperes for size 00 starters to as much as 240 amperes for size 6 starters. In larger alternating current contacts, each phase uses a three-element set of contacts. As shown in Figure 10–12, there are two stationary contacts, one each for the line and the load, and there is a moving contact with no outside connections that bridges the stationary contacts. The break at two points divides the voltage in half at each break because the circuit is in series. This reduces the arc damage to the point surfaces. It is this second category of high current power relays and motor controllers that concern us here.

All current-carrying contacts deteriorate in use. There is always some resistance between contact points. Consequently, following Ohm's law, when a current flows between the point surfaces, there is a voltage difference. This voltage difference will pit the contact surfaces. When the contacts are opened, the voltage difference dramatically rises, causing an increased deterioration of the surfaces. At some time in their service life, contact points must be tested to verify their suitability for continued use.

Toward the end of any contact's service life, a digital voltmeter with a 0.01 resolution placed across the points will show a small voltage reading. This is verification that the contacts are already in a failure mode. Contact points creating this measurable voltage drop are *already* causing increased current draw and performance problems in the equipment.

In the past, plant electricians waited until high-power relays and motor contactor points were in a failure mode before replacing them. Today, with an increasing need to control total plant power quality for transient noise and harmonics, heavy-current-carrying contactors need closer monitoring. It is very possible that a contactor on a large motor could show zero volt across the contacts as read on a DMM, and yet be the source of erratic control in another piece of equipment nearby.

The FLUKE ScopeMeter can be used to test motor contactor and relay points well before the end of their service life. Because the ScopeMeter can display a millivolt trace of *the voltage differential be-*

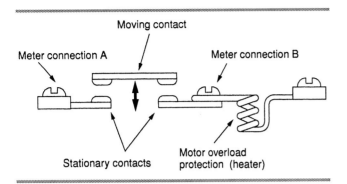

Figure 10-12 A drawing of a three-element contact. The correct locations to take voltage readings when testing the contacts for voltage drop are identified. The overload protection heaters cannot be included in the test.

increase—particularly when overvoltage spikes are present—continued service should be questioned. (See Figure 10-13B.) Also remember that other conditions could contribute to the problem, such as contamination between the armature and coil steel, an improperly seated armature, weak springs, or even current fluctuations in the line.

The allowable voltage fluctuation between contact surfaces will depend on both experience and the

tween the operating contacts, the display gives an accurate picture of their condition while in service long before a voltmeter would show point failure. Because the test is done under operating load, the condition of the contact springs and the seating of the armature itself are also included in the results.

The test shown in Figure 10-13 was done on an alternating current motor contactor operating at 480 volts. A more accurate reading is obtained if the motor contactor or relay has been in operation long enough to reach normal operating temperatures. Cold contacts will give higher-than-normal readings. The test probes were connected between the ScopeMeter's input A and the common input. *Alligator* or *hook* clips were used to attach the leads directly to the stationary contact points of the contactor. (Refer back to Figure 10-12.) The heaters of a motor contactor must not be included in the circuit. The heater coils will show a significantly greater voltage drop than the contact surfaces of the points and will mask the voltage loss directly across the points. A fully loaded heater may have a voltage drop of tenths of a volt while the desired measurement across the face of the contact surfaces will be in hundredths of a volt. The ScopeMeter will automatically range to the larger reading and the voltage drop across the contacts will become too small to read on the plot. The test is repeated for each contact set.

This is essentially a comparative test rather than one that has absolute values. Contact voltage values that are drifting between 1 and 3 millivolts in full-load operation should normally be considered acceptable. (See Figure 10-13A.) As the voltage values

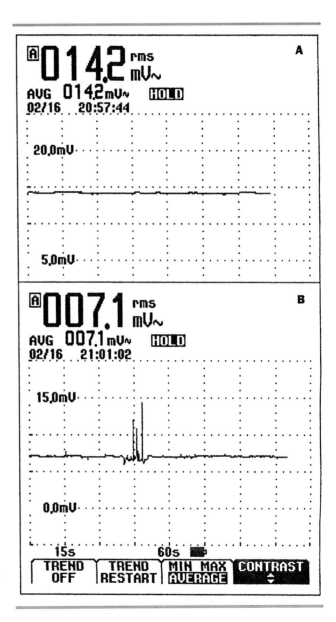

Figure 10-13 Each set of a motor controller's contacts were compared under identical operating conditions. (A) The test showed an acceptable trend plot for the phase A contacts under load (top). (B) Phase C of the same contactor under load (bottom) showed voltage spiking. The trend plot is an early indication that the contacts are failing.

type of equipment in service. If the contactor is on a line-voltage started alternating current motor at the end of an isolated branch circuit, there would be greater latitude than if solid-state motor drives were in close proximity where noise and harmonics were a concern. If the contactor was controlling a state-of-the-art production welding machine with solid-state controls, very little deviation would be allowed before weld quality deteriorated. Contractors on solid-state controlled welding machines showing 20–millivolt spikes should be evaluated for new contact kits.

It is impossible to arbitrarily state millivolt values that indicate *good* or *bad* contact points. The author contacted a research engineer at a leading electrical equipment manufacturer and was told essentially the same thing; they had done tests but could not give generalized millivolt values for *bad* contacts because the upper limit will vary with contactor size and type. This is not a test that would be done frequently. However, you may find that it proves to be a valuable test on specific equipment you service. At that point, your own experience and observation will be your reference point for evaluating contact failure.

You learn something of value in studying the traces across loaded contacts. Motor contactor or relay contacts under load are not acting as a solid conductor. The fact that the voltage is constantly drifting means that the points are analogous to a moving slip ring and brush rather than a solid busbar. Contact points in a relay or motor contactor make a **dynamic** rather than **static connection**. They are always arcing and burning even when fully closed. Look carefully at Figure 10–13A. Even good contacts do not produce an unbroken trace line. Consequently, even though opening and closing are the hardest parts of their workload, relay and motor controller contacts deteriorate even when they remain fully closed.

Trend Plot Example #4: Monitoring Mechanical Functions

Electrical power consists of two components: voltage and current. Consequently, electrical power is defined as volt-amperes or **watts (w)**. For most permanent installations in a manufacturing plant, you can assume that the voltage supplied to a given motor will remain constant. That is, from one test to another, the voltage value on the motor's branch circuit will not appreciably change. To the degree that this is true, you can measure the current of a motor-driven process to monitor the mechanical condition of all equipment and materials that the motor is driving. This is particularly true when you have permanent records that allow comparison with results of previous tests made on the same equipment.

In a pipe manufacturing plant, two identical 20–horsepower motors are parallel-mounted for a cement mortar coating process. These two motors are fed from a common branch circuit. Each motor shaft is direct-coupled to an arbor with a resilient rubber drum mounted on the arbor output shaft. These two rubber drums are adjusted so that they lightly touch each other as they rotate at 1750 rpm. In a housing to the rear of the rotating drums, an auger feeds a mortar mix into the "touch" point between the drums, resulting in a high-velocity mortar spray.

The mechanical equipment in this installation is not complicated, yet it encounters hard service and creates serious quality assurance problems in the finished product when improper adjustments in the rubber cylinders, bearing failures, or shaft coupling misalignment occurs. In the past, the only definitive test performed on the assembly before tear-down was done with a hand-held tachometer on the arbor shaft when it was running without load. However, by the time an *unloaded* motor and arbor assembly evidence noticeable speed loss, the mechanical assembly has already experienced catastrophic failure.

The maintenance department saw the need to develop better diagnostic evaluations *before* taking the mortar coating machine out of production. In effect, this was an attempt to do on-line *mechanical* troubleshooting with *electrical* test equipment. As a first step, the maintenance team discovered that an averaging clamp-on ammeter used while the motors were under load gave a better indication of system condition than no-load shaft speed. With the system in full operation, a current reading was taken on each motor phase after allowing the averaging meter to stabilize. An averaging meter will always show a false reading on the first average reset because the clamp is open for a portion of the averaging time period. In order to standardize the test, the average given on the third meter update was recorded. The three current values for each motor line were added and then divided by three for the average current of that motor under load. When the motors and all mechanical assemblies were running normally, the average current values between the two motors were similar with only a small, allowable deviation. If, on the other hand, one average motor current value was high, maintenance personnel knew that a mechanical or adjustment problem of some sort was present.

However, a second, and very useful, monitoring procedure developed with a ScopeMeter trend plot. Two standards for each motor and arbor assembly were established. The first standard is shown in Figure 10–14A when the motor is running without load. This trend plot primarily establishes a baseline, though it may also give indication of initial mechanical problems if a shaft coupling is misaligned or bearings are binding. The most useful trend plot is shown in Figure 10–14B. This second plot shows the power output of the motor over time while it is operating under full-load. (Note that the operation of the feed auger can actually be identified.) Of greatest significance, however, is the identification of an initial mechanical problem before it causes catastrophic failure. Bearings in early failure may load the motor considerably more when the process is in full operation. Bearing failures may indicate a higher power loading on one motor than the other and will be evident when the plots are compared. A severe bearing failure might also bias the plot so that decreases in load will not be as effective in decreasing motor current, so that the plot will become smoother, though elevated.

The unloaded motor and arbor assembly runs at 7.3 amperes as shown in Figure 10–14A. (The decimal point is moved one place to the right because of the 10-to-1 ratio of the current probe.) The 7.3 amperes can be subtracted from the current shown in Figure 10–14B, giving the real-time current load produced by the mortar feeding through the spinning drums. Because the feed auger is not conveying a perfectly uniform load, the trace of Figure 10–14B rises and falls. Abnormalities in the arbor bearings can now be defined with a similar record from previous performance. Since Figure 10–14A is essentially flat in normal operation, any *uniform* irregularities become a clear indication of mechanical problems. *Random* irregularities could come from poor electri-

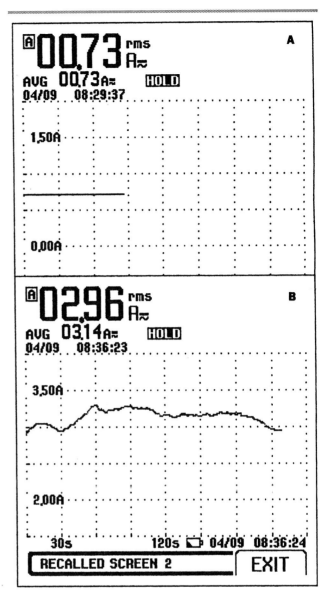

Figure 10–14 (A) A current plot over 45 seconds of an unloaded motor (top). (B) A current plot over 3½ minutes on the same motor running under load (bottom).

cal contact in the motor controller or voltage fluctuations. Say, for instance, that the plot becomes wider because the current draw is oscillating between 7.3 and 8 amperes. Because of the very narrow scale divisions for time in the plot mode, the oscillation would produce a wide envelope between 7.3 and 8 amperes rather than a narrow line. The ScopeMeter can be switched to a simple waveform reading and determine if there is an identifiable cyclical pattern. Figure 10–15 is a computer-enhanced trace of this motor in normal operation with no apparent bearing or alignment problems. This same waveform could be used as follows: If the amperage rises and falls once per revolution, it may indicate a problem with the coupling alignment or balance of the rotating assembly. If there are multiple spikes per revolution in a regular pattern, it may indicate a bearing problem in either the motor or the arbor. The cycles-per-revolution can be determined from the nameplate motor speed in relationship to the 60 hertz power. Since only one power line is being used for the current probe, the motor is considered to be single-phase.

Similar diagnostic work could be done in comparison to the working trend plot of Figure 10–14B. Since bearing problems in the arbor are greatly exaggerated under load, a trace could be started with the feed auger running, and continued as the feed process was stopped. The trace would be watched until the motor was running without load. This last trace would give further indication of bearing condition, particularly if compared with a trace from the parallel motor system operating under the same load.

The objective is to increase production by reducing downtime. In the past, this mortar coating machine was often shut down when the operator sensed a change in sound or that a motor was running hot. The motors—but not the couplings—are sufficiently exposed so that the operator's hands can be placed against the side of either motor. After the motor speeds were verified with the tachometer, mechanical evaluation was done if merited. This required removal of heavy protective shielding, disconnecting of the couplings, and verification of the bearing endplay. Many times the mortar coating machine was reassembled after the maintenance crew decided that it was running acceptably, or that future work would be scheduled at a more opportune time. The ScopeMeter monitoring of this equipment should become a useful procedure in reducing lost production.

Power Quality

In Chapter 9, you saw specialized instruments used to analyze plant power quality. This work can also be done with an oscilloscope. The ScopeMeter model demonstrated in this chapter does not have the capability of a full-power analyzer. The more expensive FLUKE hand-held oscilloscopes such as the ScopeMeter Series B meters can be used as full-power analyzers. Nonetheless, as a first step in plant power quality evaluation, the ScopeMeter can be used to indicate the presence of harmonics and transients. Figure 10–16 shows the waveform produced on the branch circuit feeding a 20-horsepower direct current drive at the 480-volt line side of an isolation transformer. Though further testing would be necessary to isolate various harmonic components, it is obvious from this simple ScopeMeter examination of the branch circuit under load that the direct current drive is producing power quality disturbances. A subsequent evaluation with a FLUKE power harmon-

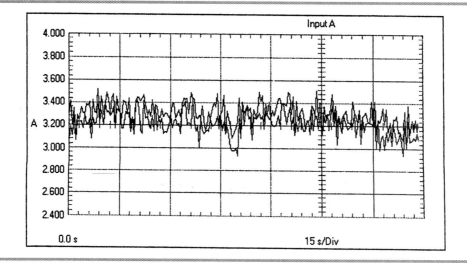

Figure 10–15 A computer-enhanced trace of the motor current could be used to show motor and arbor bearing problems. This trace indicates no abnormalities.

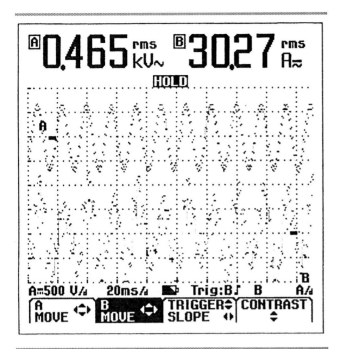

Figure 10-16 The power quality of a branch circuit with a DC variable speed drive running under load. The voltage plot (A) is symmetrical; the current plot (B) indicates heavy amperage switching loads.

ics analyzer 41B showed some third harmonic and heavy fifth harmonic components in the latter trace.

Using an Oscilloscope with a Manufacturer's Service Center

This book focuses on reducing costly plant downtime. In many cases, bringing in a field representative to troubleshoot and repair a specialized piece of equipment is often the wisest choice in protecting production schedules. This may be true *provided that the technician is readily available or that acceptable production can be maintained until the technician can arrange a visit.* But this is not always possible.

With proper test equipment, there is often an alternative. As the plant electrician, you may not have the specialized training necessary to adequately troubleshoot a catastrophic electrical failure in variable speed drives, welding equipment, or other process equipment. Proper recording of the electrical output, however, may be all that is needed for the equipment manufacturer's technical assistance team to diagnose the problem.

Most equipment manufacturers provide technical assistance by phone. This may be through a regional service center, though you may also be put in direct contact with manufacturer's engineers. The helpful and knowledgeable assistance these engineers and technicians are able to give is impressive. Be prudent, however. Genuine emergencies and pertinent technical questions are always welcomed, but avoid frivolous questions that may already be answered in their equipment manuals.

Say, for example, that you lose production because a variable speed drive shuts down. Assume that you either programmed the drive during its initial installation or that you have familiarized yourself with the user's manual and the program being used. Consequently, your first step will be to bring up error codes and determine what fault shut down the drive. The error code may direct you to something like "LOST CONTROL POWER." In turn, you may check fuses on the drive itself and find a blown fuse. (This is an example from the author's personal experience. The error code resulted from an operator shutting down the entire machine by tripping the 400-ampere main circuit-breaker with the drive energized. In spite of an isolation transformer between the line and drive, the surge blew an ATQ 15/100 fuse—rated at 0.15 ampere—protecting the logic board.) On the other hand, the control error may take you to something far more serious, indicating an improper power output.

For the sake of this example, assume that you can safely reset the fault and bring the unloaded motor to half speed, even though it is running poorly. At this point, you can record an instantaneous waveform of both the voltage and current output on your ScopeMeter. If it is a direct current drive, you will want to record both armature and field waveforms. You may also want to record the input alternating current waveforms. Assuming that you have the available computer equipment and software to do so, you can then print the graph information from the ScopeMeter. By sending this information to a manufacturer's service representative, he or she may very likely be able to diagnose the problem. In all probability, this will involve several phone conversations. The technician may have specific questions that will require voltage measurements to answer.

You may—or may not—be able to get the machine

back into immediate production. Repair may require a replacement part from a local distributor, or ordering something by overnight air delivery. Nonetheless, by making the waveform information available to a factory technician—and by replacing parts yourself—you may be able to get the equipment back into production much quicker than by waiting for outside service. When it is not practical or safe for you to do the repair work yourself, you may still be able to reduce lost production time by providing the factory service technician with information so that the necessary replacement parts can be ordered ahead of time.

There is also a very positive benefit to you, the electrician, in this procedure of working with manufacturers' representatives. It is an excellent opportunity to build your general knowledge of a particular type of equipment, plus the experience will better familiarize you with shortcuts to getting equipment in your plant back into production in the future.

The Electrician and Quality Assurance

The alert troubleshooting electrician can find opportunity to do much more than "fix" broken equipment. In Figures 10-6 and 10-7, you saw traces from welding machines that had faulty rectification. Interestingly, these traces were both made on welding machines that were in production; the operators were not complaining of any machine problems. In both cases, routine monitoring of the welding machines by the electrician was the first indication that potential problems existed. Though a good welding operator could still pass X-ray tests with these machines, they were imperceptibly becoming harder to control. At some point, they would become the source of poor welds for less-experienced operators.

This electrician made a major contribution to the quality assurance program in the welding production plant. This testing is aimed at more that mere diagnosis of failed welding machines. The electrician has now become the person who will detect the earliest deviation in welding machine *performance*. This electrician may suggest to management that a periodic test of each machine be conducted with a printout as shown in Figure 10-5 or Figure 10-6. At each test period, the current printout will be compared with the previous printout. In this way, very early machine problems can be corrected before weld quality falls below specification and causes expensive rework in the shop or failures in the field.

Because of either the type of rectification or the control boards used, perfectly symmetrical waveforms may not always be produced on the output side of a welding machine. The welding process has the ability to average modified current and voltage amplitudes and produce a satisfactory weld. What will be important, however, is the development of a trend. Each new machine could be subjected to a standard test with a fixed load at the time of initial installation. Thereafter, at regular intervals, the machine would again be tested with this standard load. The electrician conducting the periodic tests is now looking for a deviation from the results of these previous tests more than having a preconceived notion of what an ideal waveform should look like. In addition, traces made on new machines may also document the need for future warranty replacement, thus saving the company repair costs.

This discussion of the value of a highly qualified troubleshooting electrician has now come full circle. You are no longer merely looking at the person who can quickly repair electrical problems in a manufacturing plant in order to restore production, though that will remain one of the highest priorities of the troubleshooting electrician. You are now looking at an electrician who can set the standard for innovative quality assurance programs by monitoring the equipment in the plant to ensure that production output and efficiency is at its optimum level. However, from personal experience, the author also notes that management must have the vision to entrust this task to the electrician capable of innovative work. Many worthwhile solutions have never been initiated simply because a manager does not understand how an electrical value like *amperage* could result in a more profitable or maintenance-free production line.

Using an Industrial Oscilloscope in Your Own Plant

Specific ScopeMeter examples were used to illustrate its use in an industrial plant. However, your plant is unique. Now *you* must take the test examples and individualize them to your own needs. There may be ScopeMeter functions you will want to use that were not covered in this chapter. Nonetheless, the two values of voltage and amperage will be invaluable in any industrial plant. With some patient testing, you will find that most equipment will produce characteristic voltage and/or amperage waveforms in normal production.

You will find that one of two meter functions will best display the information for your use. In some cases, the instantaneous waveform itself is the most useful. In others, a trend plot will give you the best information. Sometimes, both are alternately useful.

In all likelihood, when you first start measuring the performance of equipment in your plant, you will not fully anticipate what you are looking for because a waveform will give you much greater information than the numerical voltage and amperage averages you have been reading from a digital meter. You will need to repeatedly watch the trace of a given machine as it is running. If at all possible, obtain the necessary software so that you can keep a record of acceptable operation for each machine. Store the waveforms and keep them on record in either your computer memory or as a printout in a notebook. You can also print directly from the ScopeMeter with interface accessories provided by FLUKE. This option does not require a computer, though it gives no permanent memory storage.

Note: You can bring the real-time display of the ScopeMeter directly to a laptop computer in the field. The computer will enhance the screen display and may allow greater precision in your testing than with the ScopeMeter alone. However, to reiterate an earlier warning, when coupling an energized meter to a computer, always use a meter with an optical interface. The ScopeMeter uses an RS232 interface with added optical isolation. With a hard-wired (phone jack) interface, transients could pass from the meter to the computer. Neither the computer, nor the electrician using the computer, are rated to withstand 480–volt transients!

When the machine you have been monitoring begins running poorly and produces substandard welds, poor batching, erratic conveyor movement, or whatever is characteristic to that equipment, again record the waveform or trace and compare the poor operation trace with the earlier trace. What has changed?

Deviation from the former trace in the voltage plot may well indicate connection problems. For example, you may discover that you have a poor connection in a fused disconnect or busbar lug. Yet, not all problems that *appear* to be electrical in nature are. They may be mechanical. Deviation from a former trace in the amperage plot of a motor-driven machine may indicate mechanical failure in bearings or gear boxes. Look for recurring amperage surges, which are cyclical and are in time with machine movement or gearbox output. If there are two or more similar motor or solenoid functions, compare output. Erratic hydraulic solenoid operation from a poor relay contact should become obvious through either a voltage or amperage plot trace on the solenoid coil.

Will your employer allow you the time to catalog performance for critical equipment in your plant? Obviously, this would be a waste of time for *every*

A DAY AT THE PLANT

My own experience may suggest how you might proceed. Even though I had used a bench oscilloscope for testing welding machine output, alternating current motor characteristics, and the like, I limited its use for fear of damaging the oscilloscope itself. When I began using the ScopeMeter, I felt much greater freedom to broadly test critical equipment in the plant. It was only after I looked at many traces from welding machines that I began to realize the importance of certain plot and waveform characteristics. It was only after I was actually cataloging a number of recorded traces for some thirty machines that I realized the significance of using the output readings for quality assurance of the welding machine itself.

motor or system in your plant. The wise employer (and plant supervisor) understands the need for improved efficiency and reduced downtime. Nonetheless, you may need to start conservatively. Choose a piece of machinery that is maintenance prone. Carefully log voltage and amperage characteristics when production is favorable. The day will come when you can take similar measurements when the equipment is failing. When your new procedure allows you to quickly diagnose and correct a long-standing problem, you will have the example you need to show the value of your careful diagnostic effort. This will be particularly true when you find a mechanical solution to what had long been presumed to be an electrical problem.

An erratic welder wire feed problem could be directly checked with a ScopeMeter. Visualize what is occurring in the remote wire feed motor drive when the wire is binding in the mechanical wire guide. The wire feed motor has a set direct current voltage for a given wire speed. However, if the wire is binding, the motor current will rise. A simple direct current plot on the wire feed drive motor should indicate when the wire feed is mechanically binding. If an erratic current plot of the wire feed drive motor corresponds with the faulty welder performance, the fault is in either the motor itself, the gear drive, or the wire guide.

You, too, will find many opportunities to quickly diagnose electrical problems, identify mechanical failures, and even establish quality assurance data by knowing the electrical waveforms of production machinery in your plant.

A DAY AT THE PLANT

In our own plant, before I began using the ScopeMeter, we encountered a critical welding problem on a pipe mill. One of two welding stations was performing satisfactorily, whereas the other was periodically producing substandard welds. Before the problem went away—because it was never really solved—we had a highly qualified factory representative watching the weld process for almost six hours. We swapped circuit boards between the two wire feed units numerous times, we changed contactors in the welding machine itself, we had a service technician load test the welder output, and we made numerous mechanical adjustments. But worst of all, we made pipe that needed expensive hand repairs. Three days later the problem went away, though no one could figure out why. Now, after using the ScopeMeter to watch the process—and having encountered the problem again—we are certain that erratic wire feed caused by binding in the mechanical wire guide was the real cause. It was never an electrical problem, even though it had all the characteristics of a circuit board failure. The problem is now correctable in an hour by changing the mechanical wire guide between the remote wire feed motor drive and the welding head. The ScopeMeter tells me it is not an electrical problem.

CHAPTER REVIEW

Oscilloscopes, which are both portable and sufficiently immune to adverse voltage effects, are now available for industrial use. These hand-held, industrial meters offer many functions that allow maintenance electricians to directly measure electrical power and events in a manufacturing facility.

Industrial testing with a hand-held oscilloscope will generally involve three test modes:

1. The meter will be used to evaluate the *instantaneous waveform* of voltage and/or amperage of equipment in operation.
2. The meter will be used to collect a *trend-plot-over-time* of voltage and/or amperage of equipment in operation.
3. The meter will be used to monitor the *power quality* of the plant's electrical system.

This information will be used in one of two applications:

1. The resulting measurements will be used for *immediate diagnostic purposes*.
2. The information will be logged so that *performance-over-time* may be monitored.

Because the current draw of electrical equipment in operation is a reflection of the condition or operation of the mechanical machinery being driven, the industrial oscilloscope may be used to monitor *mechanical* functions in process machinery or *quality assurance* standards in the end product.

THINKING THROUGH THE TEXT

1. Describe a *waveform*.

    a. What type of information does a *waveform* give?

    b. Give an example of how *waveform* information might be used in industrial plant troubleshooting.

2. Describe a *trend plot*.

    a. What type of information does a *trend plot* give?

    b. Give an example of how *trend plot* information might be used in industrial plant troubleshooting.

3. While recording voltage information, an industrial oscilloscope must be connected to line voltage.

    a. What personal safety precautions must you take while using this test equipment?

    b. What safety precautions should you exercise to protect the oscilloscope itself from overvoltage damage?

CHAPTER 11

Expanding On-Line Troubleshooting Applications

OBJECTIVES

After completing this chapter, you should be able to:
- **Apply basic on-line troubleshooting skills to both highly complex circuits and simple circuits.**
- **Proceed with troubleshooting when you have no ladder diagram information.**
- **Extend your troubleshooting awareness to the industrial plant's power quality.**
- **Include fuse technology as a preventative measure in your troubleshooting work.**

To this point in the book, you have been working with line segments of a moderately sized ladder diagram. If you are currently servicing equipment in an industrial plant, you will see many electrical circuit diagrams that are similar to the one you are studying in this book. They will be similar in their complexity and types of circuits.

However, if you have grasped the significance of the troubleshooting information you have studied so far, you will realize that increased size and complexity in a ladder diagram is not a question of new kinds of information. Rather, it is simply a matter of adding more of the same information you have already been dealing with. Conversely, in the case of a more simplified diagram, it will be a matter of reducing the number of circuits, not a matter of changing the type of presentation.

If you understand the basic electrical symbols used on a diagram, its complexity or simplicity is merely a matter of dealing with more or fewer circuit functions. It is not a problem of dealing with new forms of information. You presently have all of the basic information necessary in order to successfully use any electrical diagram you might encounter.

READING COMPLEX ELECTRICAL DIAGRAMS

How will you respond if you are given a complex set of drawings for your next troubleshooting task? If you carefully look at the drawings, you will realize that the complexity does not come from the need to understand new symbols or drawing formats. The complexity is merely in the number of circuits you will be dealing with. If you logically work through the new material with the information you already know, you should have little increased difficulty in understanding the diagram.

You will still use your basic ladder diagram reading skills with a complex diagram. As in any ladder diagram, each circuit will be represented as a line running from left to right on the drawing. However, you will see some differences.

1. *More circuit areas will be involved.* In all likelihood, the more complex the diagram, the more you will find widely displaced electrical functions represented on a single circuit line. A simple diagram may have two or three relay or switching points on a line that are physically close to each other in the actual electrical panel. As the electrical diagram grows, you will find functions on a single line that represent widely separated physical locations for switches and relays, as well as the involvement of numerous electrical circuits and functions. Consequently, you will need to be aware of many more parts of the circuit to do your basic troubleshooting. Nonetheless, the testing procedure is the same.

2. *Circuits will be symbolically divided.* As the circuit grows, it becomes increasingly difficult to keep related circuits close to each other on the drawing. For this reason, you will see a greater use of separated circuits designated with arrows and wire numbers. Any circuit malfunction will demand more care on your part so that you do not overlook continuations of circuits in other areas of the diagram. That is particularly true of parallel circuits that have an influence on the circuit you are testing. There will also be a greater use of widely separated mechanically connected functions, such as double-throw limit switches and multifunction push buttons and switches.

3. *There may be a greater use of specialized symbols.* Though not always so, in many cases, a more complex diagram will also include more specialized symbols. Thus, you may have various thermal, pressure, or other specialized limit switches indicated on the diagram. In addition, there may be a greater use of solid-state equipment and their related symbols.

The basic guideline in working with complex diagrams is to isolate the test areas into small segments of the diagram. Once you have narrowed the work you will be doing to a manageable area on the diagram, the testing procedures will be the same as those you have used with simpler diagrams.

READING SIMPLE WIRING DIAGRAMS

You will often find wiring diagrams in equipment with limited electrical circuits. This is particularly true of self-contained motor and motor overload circuits. For example, in Figure 3–1 you have a wiring diagram for a refrigeration compressor motor circuit. This type of diagram is often attached inside the terminal cover. You were given an explanation of wiring diagrams, as opposed to ladder or line diagrams, in Chapter 3.

Reading this diagram should not give you any difficulty. The actual symbols are essentially the same as those used in the ladder diagram. However, the wiring diagram will arrange the drawing according to the physical layout of the electrical equipment it represents. This is in contrast to the ladder diagram, which will lay all of the circuit components on a horizontal line without regard to their actual physical location in the equipment.

For the purpose of troubleshooting, you can use this diagram in the same way you have used the ladder diagram. If you are using on-line troubleshooting techniques, you would trace through the live circuit checking for an open conductor. If you have taken the equipment off-line, you would check for open conductors by doing resistance (or conventional continuity) measurements. In any case, the testing procedure will be much like you have done elsewhere in this text.

You may also find that more is left to your assumed knowledge with less complex equipment. For example, in the refrigeration wire diagram used in an earlier example, there may be no additional circuit information given even though a motor relay, refrigeration cut-out switches, fuses, and a control switch are in the actual circuit. If you are troubleshooting the unit, you may be required to do some basic testing (visual and otherwise) to determine parts of the circuit not shown on drawings inside of the equipment. Of course, if you can obtain the service manual, complete diagrams should be available.

Troubleshooting without Diagrams

Someday you will open a panel full of relays and wires and be asked to "fix it," but you will have no ladder diagram. With the information in this section, you can eventually complete your troubleshooting work even though it will take more time.

There is no single formula to use when troubleshooting without a diagram. Nonetheless, there are techniques you can use to greatly reduce the confusion. In this section, we will show you one way to get started. From your own experience, you will find other workable procedures. Do not be afraid to experiment. Since you are troubleshooting without a diagram, we will explain what is being done. You will need to carefully visualize the process because there are no specific references to wire and identification numbers.

Your Initial Testing Sequence

Now your work begins. You have been called to a manufacturing plant to check a nonfunctioning ejector system on a plastic injection molding machine. The company has no ladder diagrams for this machine. The machine has been shut down, and they want you to get it back on-line. To simplify the example, you will get through the preliminary steps of your investigation without recording it here. Finally, however, you concur that the ejector system is truly at fault as you have been told. For simplicity's sake, as in the example of the ejector system in Chapter 8, you can verify the hydraulic solenoids with the manual ejection push button. Fortunately, the present electrical problem is confined to the automatic functions. Your first step will be to isolate the electrical device or component (a solenoid in this case) that is not functioning. That will be the starting point for your subsequent testing.

Locating a Non-Functioning Solenoid

There are three ways in which you might locate the hydraulic solenoid that activates the ejection system. The first—and simplest—is to find the location from some information source other than the missing electrical drawings. The operator's handbook may list each of the solenoids and its function, there may be hydraulic diagrams available that give the information, or the solenoids themselves may have a descriptive tag on them.

The second way to locate the solenoid is through its operation. In this case, since you can use the manual push button, you might have an assistant cycle the ejector while you look for the physical location of the solenoid. You should be able to hear or feel the solenoid "click" when it is cycled. If it is a four-way valve with two solenoids, you can even determine which solenoid is activating by lightly holding a small Phillips screwdriver against the valve's manual actuating pin. In almost every case, a hydraulic solenoid valve will have a push rod that is used to manually cycle the solenoid. If the screwdriver is drawn toward the valve when it activates, the solenoid facing the screwdriver is the one cycling. If the screwdriver is forced out of the solenoid, it is the solenoid on the opposite side of the valve which is activating. Be extremely careful, however, that you do not unintentionally cycle the valve with this procedure.

The final way to locate the solenoid is the most difficult and demands the most care for the sake of safety. In this case, you would trace the ejector cylinder's hydraulic lines back to a hydraulic valve or manifold block. This may identify the valve and solenoid or it may identify two or three possibilities. You can now verify the correct valve by manually pushing the actuating pin on the solenoid. By watching the ejector system, you can verify the solenoid by its function. Needless to say, this procedure requires special care, since it is done with the hydraulic pumps running. The machine must be set so that the actuation of the wrong solenoid will not damage the machine or injure personnel. You must also remember that you have bypassed all limit switches, stroke safeties, and electrical interlocks by manually activating the solenoid. Be certain that you know what you are doing before attempting this test procedure.

Electrical Testing from the Known Electrical Device or Component

The purpose of your testing is to work back from the known, nonfunctioning electrical device or component to find the circuit problem. You should be able to remove the solenoid cover and identify the wire number of the circuit feeding that solenoid. After testing the solenoid for continuity, testing from this wire number is the next step in your troubleshooting sequence. *Unless there is a specific need for measuring a voltage value, these tests are done* after appropriately locking out the disconnects to the equipment.

What do you do if there are no wire identification numbers on the solenoid wires? Wire identification tags are often missing when wires have been reconnected a number of times. However, if you do some careful continuity testing, you can find your wire number relatively easily. Disconnect both leads from the solenoid. Attach a test wire that is long enough to reach the electrical panel to one of the two wires under the solenoid cover. Use only the supply wires from the panel, not the solenoid leads. First, test the grounded common. Change the test lead on the two wires feeding the solenoid until you find continuity. You can now identify that wire as the common. Unless the common wire itself is damaged, it is of no value to you as a part of the test, so you should identify it by reconnecting it to the solenoid coil, and begin your test with the control wire. Connect your test lead to the remaining solenoid supply wire and again use your meter to find continuity with a numbered wire in the electrical panel. By touching each of the wires on the terminal block(s), you should find one that shows continuity. That is, of course, the wire—and the identification number—that is controlling the unidentified solenoid. If none of the terminal block numbers indicate continuity, you may need to test in other parts of the machine as described above.

Figure 11-1 Using a wire identification instrument to locate solenoid control wires in the panel. This procedure may be used to identify circuits if wire labels are missing. This testing is always done on de-energized circuits. *Test instrument courtesy of Greenlee Textron.*

If you have a wire sorting tool as shown in Figure 11-1, you can simplify this portion of wire identification if there are multiple wires that have lost their identification numbers. On the other hand, there may be clues in the solenoid itself. Many times, the control wires to a four-way solenoid are sequentially numbered. On lines 9 and 10 in Figure 11-2, the mold height smaller solenoid is controlled

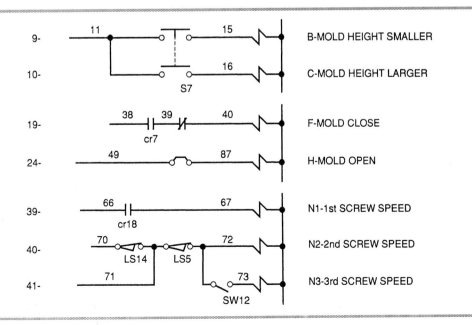

Figure 11-2 Wire numbers for hydraulic solenoids are often sequential. This can help you identify wires with lost numbers.

by wire 15, and mold height larger is controlled by wire 16. In some cases, the wire numbers for solenoids in similar locations of the machine are at least close. Three screw speeds on lines 39, 40, and 41 are controlled by wire numbers 67, 72, and 73. However, this is not always the case. If you look on lines 19 and 24, you see mold close as wire 40 and mold open as wire 87. Because of an apparent machine modification, there is a jumper on line 24 that changes wire 49 to 87. Nonetheless, if only one wire number is missing in the solenoid terminal box, the other number may give you an idea where to start testing on the terminal block for the unidentified wire.

As an added precaution, when you have identified what you believe to be the correct wire number in the panel, check all other incoming wires to make certain that the wire from the solenoid is not common to any other wire numbers. These tests will usually be confined to terminal blocks if the panel uses them.

Testing toward the Fault

You now have a wire number that you know should lead to the solenoid that is not functioning. You might wish to verify this with a voltage reading on the wire you have just identified by manually cycling the function or running the machine until it stalls. The reading should be zero volt if you have identified the correct wire and if the solenoid coil is not the cause of the malfunction. You will now begin your testing from this wire number back to the left-hand side of the circuit.

Actual Troubleshooting Procedures

There would be times during the testing in which you could use on-line troubleshooting with a live circuit. However, when you are testing limit switches that are next to moving equipment, extreme caution must be exercised lest the machine cycle. Generally, if you are tracing a circuit through the actual machine, it is recommended that you completely de-energize the prime movers (motors). If you can locate the necessary wire numbers in the electrical panel, you might wish to stall the machine and use on-line procedures, provided you can satisfy all necessary safety precautions.

Your present testing objective is to identify each end of the numbered wires in the ejector solenoid circuit. If, for example, you determine that the wire to the solenoid is wire 55, you need to trace this wire number to some controlling device on the machine or in the control panel. If you cannot find wire 55 coming into a panel terminal strip, you may then need to trace it through a limit switch or similar device on the machine. However, because this wire is a solenoid, it will probably return to the control panel. You must now find this same wire number on a control point in a relay contact or a switch of some kind that must be inspected. If wire 55 terminates on a relay, you must trace across the relay contact to identify the next wire number. You will, of course, verify the condition of each control point (relay contacts, limit switch, etc.) in the circuit as you progress. If wire 55 terminates on a relay contact that closes across wire 54, then your next step will be to locate the other end of wire 54, check that point for conductivity, and proceed to the next wire number. If, for instance, you find more than a single wire at any terminal, you know that you need to trace multiple termination points. If two wires are labeled as 55 on the first relay you find, you must trace both 55 wires to their sources and perform tests on each of those circuits. As you work, you will need to draw the developing diagram and keep notes.

This step-by-step testing will take you to many parts of the machine. A wire number may have only one termination point in the panel. It may be obvious that the wire is routed from the panel terminal block to a conduit exiting the panel. You will then need to trace that wire to a control point such as a panel switch or limit switch. You will need to open the panel and switch covers to determine wire numbers and locations. On the other hand, you may be able to knowledgeably guess and eliminate some steps. Wire numbers across a single switch are often sequential. Thus, if wire 55 leaves the panel, and wire 56 returns to the panel, you might check across these wires to see if they are common. A circuit often has wire numbers that are close in numerical sequence. With the covers off, you can check that specific switch for continuity, moving the switch to check its operation in both its open and closed positions.

When you are tracing wires outside of the electrical panel, you can use conduits and wire channels as location indicators. If you are trying to locate a termination point for an ejector-related circuit wire, you would first trace the wire in the panel to a specific conduit exiting the panel. That conduit then goes to a given part of the machine, either directly or via a wire channel. If any of the limit switches on the ejector system are fed by that conduit or wire channel, they will be the first switches to check for the given wire numbers.

As you locate and record the electrical diagram for the circuit you are testing, you should draw the diagram with the appropriate electrical symbols. Your troubleshooting techniques for the circuit devices will be much the same as described in other sections of the book. The difference, however, would come only in the thoroughness of your testing as you move from section to section. Since it would take considerable time to draw the entire diagram, it would be a better use of time to completely check each area of the circuit as you move through it, hoping that you would find the problem early. Unless you are attempting to re-create the electrical diagram for future reference, there is little value in continuing to draw the diagram after you find the system failure.

Not all of your troubleshooting will be neat and clean. Someday you will open panels like the ones shown in Figure 11–3 and Figure 11–4. Wire numbers will be missing. Changes have been made by electricians before you, without making notations on the prints. Worst of all, the prints may not exist! And it is your job to find out what is wrong and get the machine back into operation.

TAKING YOUR TROUBLESHOOTING SKILL TO THE PLANT

Reducing lost production time is your objective in developing troubleshooting skills. Throughout this book, both your troubleshooting techniques and your use of appropriate tools as a means to this end have been emphasized.

Sometimes, however, you can optimize your troubleshooting effectiveness by working in areas that extend considerably beyond the equipment you are servicing. Briefly consider two subjects that are often neglected: plant power quality and appropriate fuses.

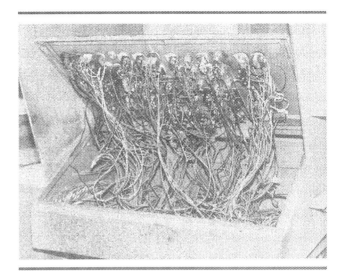

Figure 11-3 This kind of wiring really does exist. Many of these wires have no number identification. This is the operator's console for a pipe mortar coating machine.

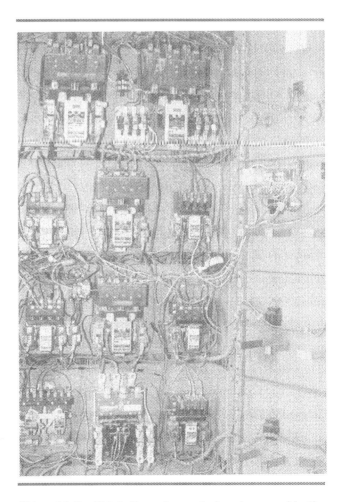

Figure 11-3 This is the motor control center served by the operator's console in Figure 11-3. There are even fewer wire identification numbers in this panel! Notice the antiquated two-pole thermal protection.

Evaluating Plant Power Quality

In Chapter 6 you were told that a 3.5 percent voltage imbalance will cause an approximate 25 percent increase in motor temperature. You may then compound your supply voltage imbalance further with harmonics, low power factor, and other "unseen" power problems.

What do you do when motors fail for unexplained reasons? How do you attempt to "fix" solid-state equipment that is erratic or does not operate properly? What do you do with a PLC-controlled assembly line that does predictably strange things several times a day? Do you simply exchange motors, make a pretext of adjusting proximity switches, and add redundant mechanical limit switches to stop machine travel? Or do you take the time to determine *why* the problems are occurring?

Evaluating the power quality of a plant may be the first step in reducing significant production losses and repair costs. In the last chapter, a single example was shown of the ScopeMeter's use in power quality evaluation. As a troubleshooting electrician you need to consider the effects of power quality disturbance on the equipment you are maintaining.

Even as the need for cleaner power increases, more and larger **nonlinear loads** are added to a manufacturing plant's power system. Most electronic loads are powered by diode-capacitor power supplies that draw current in short pulses. Each new computer, copier, electronic ballast, or variable speed motor drive adds more harmonics to the power system. Today, most production plant power supply problems originate from within the plant rather than from the outside service entrance.

The FLUKE 43 shown in Figure 11–5 is an advanced power quality analyzer with multimeter, ammeter, and oscilloscope functions. It can accurately measure parameters such as true power (watts), **apparent power (VA)**, **reactive power (VAR)**, and the power factor. Its monitoring functions help track intermittent problems and power system performance for up to 16 days. In its transient monitoring mode, the FLUKE 43 can capture and save up to forty voltage transients, which are stamped with the time and date, allowing comparison with other equipment in

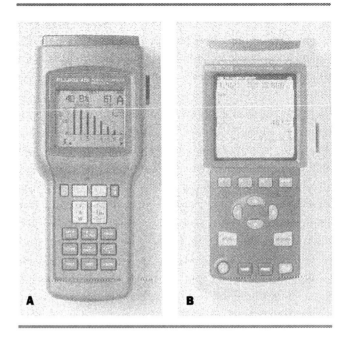

Figure 11-5   The FLUKE 41B (A) and 43 (B) Power Quality Analyzers. *Courtesy of FLUKE Corporation*

use. A recording technique called "Sags & Swells Mode" can monitor up to 24 hours and indicate fluctuations as short as a single line cycle. It interfaces with a computer to analyze and store the recorded information.

Other less expensive power quality analyzing equipment is also available. The FLUKE 41B is capable of displaying waveforms, analyzing harmonics, and displaying many of the same values as the FLUKE 43. However, it does not have the ability to record power quality over extended periods of time. The FLUKE 41B is also shown in Figure 11–5.

There may be times when you would test for single values with a dedicated meter. As shown in Figure 11–6, watt and power factor meters with a single function are also available.

This brief section is not intended to show you how to test for power quality problems or how to make corrections when you find them. It is simply to remind you that your best troubleshooting efforts on balky machinery with electronic controls or variable speed motor drives may be largely wasted if you are fighting supply line power distortion. Today's maintenance electrician will be increasingly expected to know what is happening in the power supply of his

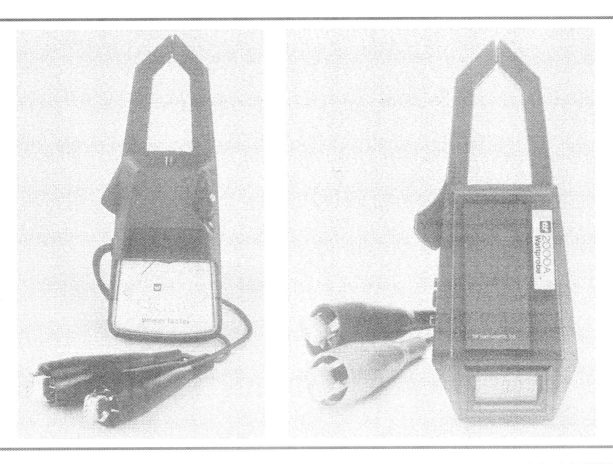

Figure 11-6 Two meters from Tif Instruments, Inc. Their TIF2300 power factor meters is show on the left; their TIF2000A watt meter is shown on the right. *Courtesy of Tif Instruments, Inc.*

or her plant. A good troubleshooting electrician will consider the incoming power to any machinery with solid-state equipment as a part of the preliminary evaluation, which was explained in Chapter 6.

Proper Fuses Reduce Lost Production

The *National Electrical Code*® (*NEC*®) is committed to the *safe* use of electrical power. *NEC*® is *not* committed to the optimum production from your plant. As we will see, a circuit-breaker is sized to provide safe protection of the plant's wiring. It does not necessarily ensure that a motor will not be damaged by overload. *NEC*® represents the *minimum* standard for safety. If you want to keep a manufacturing plant at optimum production, it is your responsibility to exceed the minimum standards of the *NEC*®. Fusing motors is a good example.

Fuses Versus Circuit-Breakers

Figure 11–7 compares the fault-current let-through of a circuit breaker and a high-**current-limiting fuse**. Underwriters Laboratories Inc. (UL) states that *current limiting* indicates that a fuse at a rated voltage will start to melt within 90° electrical and will clear the circuit within 180° electrical. This means that a properly sized fuse on a 60–hertz installation will open in 0.00833 second. On the other hand, a mechanical circuit-breaker on the same circuit will take from two or three cycles to as long as 17 cycles (0.28322 second) to open.

Overcurrent damage to equipment results from both heat and mechanical movement from high induction forces. Whenever motor windings or other electrical conductors encounter currents higher than what they were designed to carry, insulation is subject to excessive heat, and the conductors themselves may actually melt. Look again at Figure 11–7. The

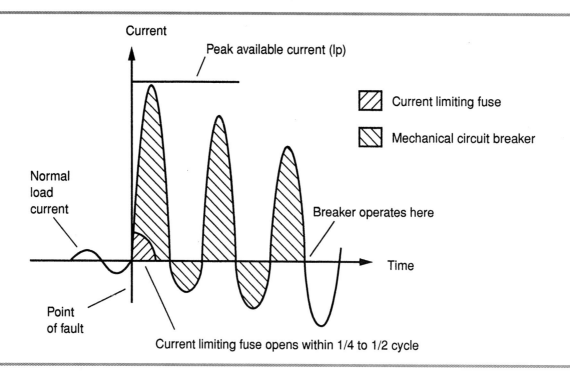

Figure 11-7 At the inspection of the fault a branch circuit can reach peak available current in double current in the absence of current limited protection. Most mechanical circuit breakers cannot respond fast enough to limit this high current in rush. The instantaneous heat Product reach temperature that destroy insulation and melt Conductors. When protected with a current limiting Fuse the Let-through current is only a function of the peak available current Julie opening the fuse in the less that one-half cycle. The shaded area in the graph represent the total heat energy absorbed by the equipment with and with either circuit-breaker or current limiting fuse protection. *Courtesy of Tif Instruments, Inc.*

actual heat produced is a direct result of the amount of current imposed on the windings or conductors. You could figure the area enclosed by the current identified as "normal load current" as being 87 percent of the rated heat capability of a 1.15 **service factor** motor running at 100 percent load. The diagram shows a fault suddenly imposed on the motor circuit. In the first example, the overcurrent protection is a high-current limiting fuse that burned within a quarter cycle. This graph indicates that the motor windings still absorbed several hundred percent of the heat they were designed to dissipate. Nonetheless, because the heat was of such short duration, the motor undoubtedly was not damaged unless the short occurred because of insulation that was already defective. On the other hand, if this same motor was protected by a mechanical circuit breaker that took only three cycles to extinguish, notice the quantity of heat imposed on the windings by the current in the shaded portion of the figure. Could this motor dissipate this much heat in its windings and not be damaged?

To answer the question properly, we need to look at *NEC®* Article 110-10, which says:

The overcurrent protective devices, the total impedance, the component short-circuit current ratings, and other characteristics of the circuit to be protected shall be selected and coordinated to permit the circuit-protective devices used to clear a fault to do so *without extensive damage to the electrical components of the circuit*. This fault shall be assumed to be either between two or more of the circuit conductors, or between any circuit conductor and the grounding conductors or enclosing metal raceway. Listed products applied in accordance with their listing shall be considered to meet the requirements of this section. (Emphasis added.)

If this motor circuit has been properly installed according to *NEC®* 110-10, the motor will not be *extensively damaged* if the equipment it is driving jams and trips a mechanical circuit-breaker. In fact, over time, the circuit-breaker may be tripped numerous times for the same reason with the motor being put back in service after each jam is cleared. However, the motor's insulation will have a cumulative heat history. Each time the insulation is subjected to excessive heat, the motor's expected service life will be reduced proportionately.

Properly installed circuit-breakers will protect the electrical equipment from *extensive* damage. On the other hand, properly sized fuses can minimize the heat history imposed on a motor resulting in *minimum* damage from an overload. We asked the question, "Could this motor dissipate this much heat in its windings and not be damaged?" The answer hinges on whether we are talking about *extensive* damage or *minimum* damage. However, the ultimate service life of this motor will be a result of its accumulated heat history. Repeated *minimum* damage will produce greater service life than merely protecting the motor from *extensive* damage.

There is more that is subject to heat damage than just the motor. The starter contacts also carry the same percentage of overload. How did they fare? If this was in a typical plant, the starter would be quickly checked and put back into service. But the contacts experienced the same accumulated overload as well. The motor starter is that much closer to causing future lost production.

And what about the circuit breaker? No one even thought to ask; it was reset and production resumed. In fact, once a circuit-breaker has been tripped, it has been compromised by the heavy load. It, too, is that much closer to failure, resulting in lost production.

Potential damage is not limited to electrical equipment. An overcurrent condition may have been caused by jammed machinery. What was happening to gearboxes, couplings, chains, belts, and other mechanical equipment during this power surge? *Everything* was overloaded. A motor coupling that fails in three months may be the result of this current overload.

Current Limiting Fuses

Not all fuses stop drawing current at the end of the first half cycle. Think back to a production welding machine. A 1,500-ampere direct current welder is drawing that current through an arc in a molten puddle, not in a direct metal-to-metal contact. The same thing can happen in a fuse cartridge. The worst example is the old "renewable" fuses, which had a replaceable link. The fuse link itself is in air and can draw an arc through many cycles as the fuse melts open. On the other hand, a current limiting fuse is filled with an insulating granular material that will smother the arc. In addition, the current limiting fuse has many segments in the fuse link so that the voltage drops quickly as it divides into a series circuit across each open link. "Renewable" fuses generally have few link segments and therefore sustain the arc at higher voltages.

High-current limiting fuses have a rating of not less than 10,000 amperes. Some are rated as high as 200,000 amperes depending on the particular fuse application. Most high-current limiting fuses used on 480 volts alternating current will be 20,000 amperes. Why does a 100–ampere fuse need a rating as high as 20,000 amperes? Because it could potentially draw very high currents under certain fault conditions. *Ferm's Fast Finder* gives the following example. Say that you supplied your plant from a 120/208–volt, 300–kilovolt-ampere, 2% Z (impedance) three-phase transformer with 20–foot, 500–**kcmil** service laterals. Kcmil represents wire diameter given in thousands (k) of circular (c) mils (1/1,000 of an inch). The service laterals come directly to the main service panel. Twenty feet from one 100–ampere fusible disconnect in the main panel you install a branch circuit panel, which you connect with #2 wire.

What magnitude of current will you have with a direct short? If there is a phase-to-phase short in the service panel, the maximum short circuit amperage will be in the range of 30,000 amperes. If there was a direct short in the branch circuit panel, the current would reach approximately 15,000 amperes. Yet the service entrance fuses are rated at 800 amperes, and the fuses for the branch panel are rated at 100 amperes. You have these high current potentials because the transformer could feed that much current into the system for a short period of time. The fuse must blow when the current exceeds its rating, but it must also *contain* the current of a direct short across the equipment to which it is connected. Refer to Figure 11–8.

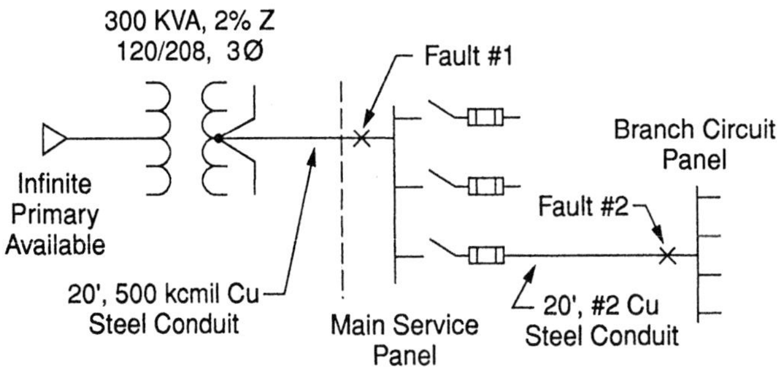

300 KVA, 2% Z
120/208, 3Ø

Infinite Primary Available

20', 500 kcmil Cu Steel Conduit

Main Service Panel

Fault #1

Branch Circuit Panel

Fault #2

20', #2 Cu Steel Conduit

Fault No. 1 Step 1. $I_{FLA} = \dfrac{KVA \times 1000}{E_{L-L} \times 1.73} = \dfrac{300 \times 1000}{208 \times 1.73} = 833\ A$

Step 2. Multiplier $= \dfrac{100}{.9^* \times Trans.\ \%\ Z} = \dfrac{100}{1.8} = 55.55$

Step 3. $I_{SCA} = 833 \times 55.55 = 46{,}273\ A$ at transformer secondary

Step 4. $f = \dfrac{1.73 \times L \times I}{C \times E_{L-L}} = \dfrac{1.73 \times 20 \times 46{,}273}{22{,}185 \times 208} = 0.347$

Step 5. $M = \dfrac{1}{1+f} = \dfrac{1}{1 + 0.347} = 0.742$ (See Table M)

Step 6. $I_{SCA} = 46{,}273 \times 0.742 = 34{,}343\ A$ at Fault No. 1

Fault No. 2 (Use I_{SCA} at Fault #1 to calculate)

Step 4. $f = \dfrac{1.73 \times 20 \times 34{,}343}{5{,}906 \times 208} = 0.968$

Step 5. $M = \dfrac{1}{1+f} = \dfrac{1}{1 + 0.968} = 0.508$ (See Table M)

Step 6. $I_{SCA} = 34{,}343 \times 0.508 = 17{,}447\ A$ at Fault No. 2

Notes: For simplicity, the motor contribution was not included.
*Transformer % Z is multiplied by .9 to establish a worst case condition.

Figure 11-8 Available fault current can reach extremely high values. Fuses must be able to contain this electrical power, which can reach level of 30,000 amperes or more. This example is take from *Ferm's Fast Finder*. For further clarification, consult any edition under the index heading, "Short Circuit Current Available, see... Short Circuit Calculation (Formula)." *Courtesy of IAEI and Littelfuse, Inc.*

Sizing Fuses to Motor Loads

Oversizing any fuse or circuit-breaker will reduce its protection of a motor. Circuit-breakers are often sized larger than a motor's running current because they are less tolerant of the high starting amperage of motor loads. With circuit-breakers, it is the motor starter's overloads that protect the motor from sustained overcurrent operation. Fuses can be sized closer to the actual allowable full current operating conditions of a particular motor operation because they have a greater selectivity of current ranges, and because dual element fuses can tolerate the high inrush current of starting.

Remember, the objective is to increase plant productivity. This not only includes fast troubleshooting, but also considers the necessary steps that will reduce future lost production. Properly installed and sized fuses may do much to increase production. It is faster to replace fuses than motors or gearboxes when equipment jams.

Fuses Can Reduce Troubleshooting Time

Figure 11–9 shows Littelfuse's Indicator™ fuses. In order to reduce cost to the end user. Littelfuse has

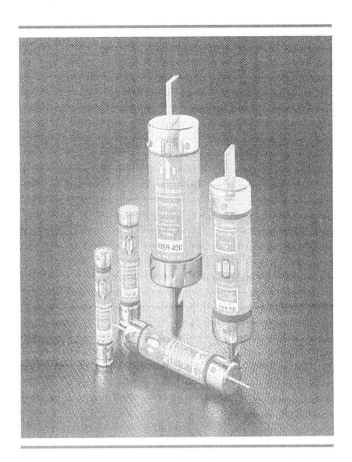

Figure 11-9 An example of Littelfuse Indicator™ fuses. These power fuses are high-current limiting and include a window for quick visual verification of the fuse condition. *Courtesy of Littelfuse, Inc.*

matched three series of fuses to specific applications: FLNR—ID fuses are used for 250-volt alternating current or less, FLSR—ID fuses are used for 600-volt alternating current or less, and IDSR fuses are used for 600-volt or less alternating or direct current. These fuses have a clear plastic window with a wound resistor in parallel with the **fuse element**. When the fuse blows, the wound element blackens the white background. These fuses can be visually verified without using a meter or disconnecting the power.

EXPANDING YOUR TROUBLESHOOTING CAPABILITIES

The major objective of this book is to give you the basic techniques for quickly troubleshooting industrial equipment with the aid of electrical (ladder) diagrams. That has been achieved by teaching you how to read electrical diagrams, giving you an introduction to some of the test equipment that is currently available, and explaining a troubleshooting technique that can greatly increase your diagnostic speed.

If you have grasped the basic principles of troubleshooting that were explained to this point, you have a good foundation for your own work in the field. It is impossible in a book of this size to cover every conceivable type of electrical testing. Furthermore, there are significant areas that have had little or no coverage at all in the book. Nonetheless, if you can use the techniques explained in this text, you will have little difficulty adapting your skills to your specific job requirements.

Distinguish yourself as a competent electrician who is fast in your troubleshooting skills. You will gain your employer's appreciation and you will increase your self-confidence as you develop professionally.

But there is more that electricians can achieve in their profession than merely installing and troubleshooting equipment.

The last two chapters of this book are aimed at broadening your skills beyond routine electrical work. If you can take it upon yourself to grow and become increasingly knowledgeable in the broad field of electrical work, you can stretch your potential far beyond your peers who are content to merely put in their day's work.

Make it your goal to become the best electrician possible!

CHAPTER REVIEW

Once you understand the basic concepts represented in electrical symbols and ladder (or wire) diagrams, the complexity or simplicity of a particular diagram is inconsequential.

In diagrams of greater complexity, which are diagrams with a larger number of circuits, you will be dealing with more widely displaced electrical functions represented in single circuit areas; you will find that the circuits are symbolically divided more throughout the drawing; and you may find a greater diversity of specialized symbols. Nonetheless, these differences should cause you little difficulty in the actual use of the diagram *if you isolate the test areas into small segments.*

A DAY AT THE PLANT

*I worked in a large plant in which fuses had been deliberately oversized with **one-time fuses** on all loads including motors. One-time fuses have no time-delay feature; they will open at the amperage rating irrespective of time duration. For budgetary reasons, I was not able to make many changes during my first year. Later, I replaced as many fuses as possible with IDSR fuses. After installing new fuses in critical machinery throughout the plant, I began to see a change.*

*First, I was being called to change fewer fuses. Properly sized **dual-element fuses** outlast oversized one-time fuses on motor and inductive loads. Dual-element fuses allow an initial overload for a short duration of time. They are used for heavy inrush currents such as motor starting. Secondly, troubleshooting routinely became simpler and faster because I could reduce fuse verification to a quick visual inspection. Finally, the real gain came in reducing the number of motor changes and damaged equipment. Management didn't fully understand what was happening. Yet I was always pleased with my contribution when called to a machine that had stopped because a jammed batch process had blown a fuse. After the blockage was removed, I could replace the fuse and put the equipment into operation without unnecessary mechanical or electrical repair. I saved money for the company, and I didn't miss lunch to replace motors!*

Equally, simple wiring diagrams should cause you no difficulty since they will also use basic electrical symbols. The wiring diagram will represent the physical layout of the circuit rather than the schematic layout of the ladder (or line) diagram. When troubleshooting simple circuits or using simple wiring diagrams, you may need to verify undesignated areas of the circuit through actual testing.

Troubleshooting without the aid of a diagram was also explained. Though you will need to develop applicable techniques for each situation, the suggestion in this chapter was to identify the known—but nonfunctioning—electrical device or component as the first step. Then, by identifying each wire number in reverse order, you can test backward toward the circuit fault. Specific suggestions were given for determining the wire numbers and the control contacts and switches in the circuit. Though on-line troubleshooting techniques may be used at certain times in this testing procedure, special cautions were given with the recommendation that far greater use of de-energized circuit testing be used. If the testing becomes extensive, you will need to draw the diagram as you work in order to properly identify and locate circuit areas for further testing.

Two areas in the physical plant were identified as meriting attention to reduce potential troubleshooting problems on machinery. The first is that of power quality. Today's electrician must pay attention to the power before it enters sensitive production equipment. The second is proper fusing, which can better protect the machinery and give visual indication of the fuse condition.

THINKING THROUGH THE TEXT

1. Briefly list and explain the three differences you might see in a more complex electrical diagram.

2. What basic guideline is given for working with complex circuits that will result in testing procedures similar to those with which you are already familiar?

3. What is the layout difference between a wiring diagram and a ladder (or line) diagram?

4. List the three ways in which a hydraulic solenoid valve for a known function may be located. What are the specific safety concerns for the third method?

5. After identifying the numbered wire connected to the final electrical device or component, what is the objective of each testing sequence?

6. When tracing a wire outside of the electrical panel, what might help you determine the locations of switches in other parts of the machine?

7. Why does a properly sized current limiting fuse give greater protection to a motor than a circuit-breaker?

CHAPTER 12

Broadening the Electrician's Horizons

OBJECTIVES

After completing this chapter, you will be aware that:

- A broad knowledge beyond mere electrical skills is mandatory for an effective troubleshooting electrician.
- Effective electrical maintenance consists of troubleshooting speed, reliability of repairs, and cost effectiveness.
- Increased trades competency will develop from reading, work experience, continuing education, and examination of failed equipment.
- The electrician has both job security and professional advancement advantages because of state electrical licensing.

THE PLACE FOR BROADER KNOWLEDGE

Helping you become more effective in your electrical troubleshooting is the underlying purpose of this book. You will achieve effectiveness in many ways. In the early chapters of this book, the information regarding diagram reading, troubleshooting procedures, and the availability of test instruments was oriented toward increasing your effectiveness as an electrician.

The techniques and skills of electrical work are important. However, unless they are balanced with in-depth understanding and knowledge, you will fail to achieve your full potential as an electrician.

Defining Effective Electrical Work

It is relatively easy to determine an electrician's effectiveness. Effective electrical maintenance is based on the interrelationship of:

1. The speed of the troubleshooting process, which is measured by total machinery downtime
2. The reliability of the diagnosis and repair, which is measured by the absence of future downtime caused by undiagnosed or poorly repaired malfunctions
3. The cost effectiveness of the completed procedure, which is measured by the ratio of the least expensive means of putting the equipment back

into service while maintaining the lowest cost resulting from lost production

Thus, you have three goals in effective electrical troubleshooting: speed, reliability, and cost effectiveness.

With a little thought, you should realize that merely learning a faster troubleshooting technique is not the entire answer to effectively getting equipment back into production. No troubleshooting technique can be set on automatic pilot. As a maintenance electrician, you need to rely on good judgment and knowledge in the way you use any technique. Your background of information on both the specific equipment you are troubleshooting and your general knowledge of the systems you are working with will have a significant bearing on how you conduct your troubleshooting procedure. The more background knowledge you have, the more apt you are to start closer to the actual problem you are looking for. More background knowledge will also increase your accuracy as you interpret your test results.

Individual Learning Patterns

Experience, of course, is an invaluable part of learning. Most of us who work in the trades "learn through our hands." Much of what we are intuitively able to do is the result of having worked through similar problems while learning from those experiences.

However, experience is not the only method of effective learning. At some point, there must be a cautious balance between two rather opposite sources of information for the tradesperson. The first source of information is actual work experience. The second source of information is the acquisition of information through study. Simply stated, this means that a well-informed electrician is going to be involved in both types of learning activities: learning through actual work experiences and expanding personal knowledge through study.

Technical Training Versus Field Experience

Technical (and particularly theoretical) information is often dismissed as being unnecessary for the field electrician. I strongly disagree! Our entire electrical profession is built around the proper application of basic laws of physics. It is true that we may not need to understand all the formulas for conductance and resistivity in order to choose the correct wire and conduit size for a given load. Nonetheless, we will certainly be better electricians if we have a general concept of what is involved in these subjects. There is more to doing a wiring job than following a table in the *National Electrical Code®* (*NEC®*). Understanding why certain wire sizes are designated should be a part of what we bring to the trade.

The need for **cross-training** is emphasized in this book. In addition to electrical knowledge, the effective electrician will have ample knowledge in many related maintenance trades and theoretical fields such as basic mechanics, power transmission, pneumatics, and hydraulics. Then, as he or she works within a specific industry, specialty subjects within that industry will also be added. This may include refrigeration, welding equipment and theory, metallurgy, chemistry, and basic thermodynamics, to name only a few. Most certainly, the effective electrician in production must become familiar with a wide range of control systems and PLC equipment.

Thus, the good electrician will have an understanding of the technical reasoning behind what is being done in the field just as the engineer must have an understanding of field procedures in order to do good design work. Or, stated in another way, I am no more impressed with the electrician who has never studied an electrical theory text than I am with the engineer who has never pulled a wire! They are both the poorer for their lack of additional knowledge or experience.

Your effectiveness as an electrical troubleshooter, which is measured by your speed, reliability, and the cost effectiveness of your work, will often be determined by your knowledge of areas outside the immediate demands of your electrical license. *Do you continue to study and learn in the field of electrical theory and practice? Do you go a step further and attempt to gain knowledge in outside—but related—fields to your work?*

The need for you to effectively relate your electrical skills to knowledge in other areas is constant. In modern industrial equipment, the electrical system is frequently interrelated with other systems.

> **A DAY AT THE PLANT**
>
> *Let me give you an example of the importance of understanding the theory of what we are doing in the field. I was troubleshooting a series of heater controllers on a large processing machine. When I got into the system, I found a strange concoction of wiring with improper lead wires used for the required J thermocouples. There were even copper wire jumpers used where the thermocouple leads were too short. Some electrician who preceded me had not done any homework on thermocouples. The same type of lead must run all the way from the controller to the thermocouple junction. The two leads for J thermocouples are iron and copper-nickel. No wonder the machine showed inaccurate readings! In spite of the time and expense required for the installation, its effectiveness was greatly reduced because the electrician didn't apply a basic knowledge of thermocouple connections.*

You will be working with electrical control components in mechanical, hydraulic, refrigeration, and pneumatic systems. In addition, you will be working with various process controls (heat, level indicators, proximity, and counter sensors, etc.) that are integrated into complex machinery. A good troubleshooter must not only understand the electrical functions being tested, but also have a good working knowledge of any other systems encountered. That working knowledge includes not only the practical applications of these other systems but their theory as well.

CHOOSING TO BROADEN YOUR KNOWLEDGE

Good electrical maintenance work, including the area of electrical troubleshooting, can become a much more challenging trade for you as an electrician if you want to take advantage of it. You can build on your skills as an electrician by systematically expanding your knowledge horizons into other technical areas. On the other hand, you can remain "just an electrician." It is your choice.

Consider three areas that offer significant potential for you as an electrician. Their value to you, however, will be determined by the kind of choices you make.

Advantages of Trade Licensing

Licensed electricians have a great advantage in the work area. For good safety reasons, the state has determined that electrical work can only be done by those who have appropriate levels of skill and knowledge. Each state has determined these knowledge and skill levels by testing and licensing. An electrician is free (legally, according to licensing law) to branch out into many other skill areas in plant maintenance. On the other hand, this same advantage is not available to others in the maintenance fields. Non-electricians (again, in terms of state electrical law) cannot do electrical work. In other words, where the job conditions merit, the electrician has many more options than anyone else in industrial maintenance. The electrician can be involved in many diverse areas of plant maintenance.

We are not suggesting that electricians take others' jobs. Yet many times we have significant latitude in our employment to broaden ourselves if we want. Though we will look at the potential advantages of working in a small shop later, this is by no means the only area in which the electrician can broaden trade skills. The large shop also has that potential. If you go into a large shop with a willingness to learn from others—and with a nonthreatening approach to their job security—you will usually be met more than halfway. Acknowledge that a fellow employee knows more about his or her trade than we do, and we will likely find a person who will be happy to be our teacher. We can also reciprocate. With no violation of the *NEC*®, we can become a source of information of electrical safety and theory. The most knowledgeable fellow-employee I ever worked with was a metal fabricator. In the four years we worked together, he taught me much. As the plant welding certification instructor, he taught and certified me as a welder. He, in turn, undoubtedly learned some information about my trade.

You have a significant advantage as a licensed electrician. In addition to the general mechanical skills in plant maintenance, there are a number of specialized skills that you could add to your repertoire: refrigeration, hydraulics, pneumatics, and even welding and metal fabrication. What you do with this opportunity, however, will depend on your personal choice. If you wish to grow, the opportunity is always there.

Allow, however, for the fact that growth will take time. After all, becoming an electrician was not something that you did overnight. This will be true with other areas of knowledge as well. However, as you begin taking on new challenges, you will see your competency increase. You will also need to work carefully (and wisely!) with lead personnel and/or management. Pace yourself so that your skills grow before you aggressively take on jobs that are beyond your ability. Damaging expensive equipment will not encourage management to entrust you with more responsibility.

The state has, in fact, given you as an electrician a decided advantage. It has given you an electrical license that allows you to work in an exclusive area without restricting your skill development in other trades. Taking advantage of that opportunity, however, is your responsibility.

Advantages of Small-Shop Employment

Every employment opportunity has its own advantages and disadvantages. Similarly, every employee is capable of using a given set of circumstances to his or her best advantage where perhaps another set of circumstances would be far less advantageous. Nonetheless, in my own experience, I have found that there are some significant benefits to working in a small shop if the employee is inclined to take advantage of the situation. There are also disadvantages, such as lower pay and longer hours as a salaried employee.

The most significant advantage in the small shop is a greater latitude of job opportunity. Typically, the small shop cannot cover each area of maintenance need with a specialist. Consequently, there is a great deal of overlap in the type of work done. If you want to broaden your work experience, this setting will give you a greater opportunity to take on other maintenance responsibilities. The employer is faced with covering as broad a range of maintenance skills as possible with limited personnel. In some cases, that may mean an attempt at covering all maintenance needs with one qualified electrician. If you are willing to work in other areas, the combination of an electrical license and your additional skills will put you in high demand. The uniqueness of the demands of the small shop will work both ways for you. Your broader skills will place you in higher demand in these employment settings. Equally, the small shop will give you an excellent place in which to develop broader skills simply because you will be the logical person to take the job irrespective of your experience in that specific maintenance area.

Job status is another consideration favoring a small shop. The advancement progression in a large plant from maintenance electrician to plant manager may represent twenty or more years of seniority and a great deal of political know-how. In a small shop, it may be the same job. The real advantage, however, comes in using the job as a basis for professional advancement. Let's use the example we just introduced. After three years of effective work in the above large plant, an electrician looking for other employment will be doing so as a low-ranking plant electrician with three years' experience even though his or her job references may be excellent. On the other hand, the same electrician who had been the sole maintenance person in a small shop may quite legitimately represent him- or herself as a plant maintenance manager with three years' experience. That qualification may give the same electrician an opportunity for a maintenance manager position in a larger company. There are many reasons why a person may choose to stay in small-shop employment, but there is also the very real opportunity for rapid professional advancement by spending time in carefully selected small-shop jobs.

Thus, you have some choices you can exercise in the type of employment you seek that can have an important bearing on your ultimate professional standing. Choosing the size of plant you wish to work in with the specific goal of broadening your knowledge and experience background can have significant benefits for you if you are careful and realistic in your planning.

Advantages of Developing Unique Skills

Your personal marketability in the employment field will depend to a large extent on the unique skills you can bring to your employer. If you can bend conduit and pull wire like any other one of a thousand fellow electricians in your city, your marketability is no better than theirs; the competition will remain high for a limited number of jobs. On the other hand, if you can acquire unique skills that set you apart to a prospective employer, your job security and opportunities for advancement will greatly increase. The focus of this book is, of course, troubleshooting. Electrical troubleshooting is certainly a skill that will greatly enhance your professional advancement. In the next few paragraphs, we want you to gain an appreciation for some cause-and-effect relationships between the development of unique skills and effective troubleshooting.

You should have gathered by now that my personal philosophy of effective troubleshooting emphasizes your knowledge of the equipment you are working on. In addition to the electrical components, this will include mechanical, hydraulic, pneumatic, and whatever other systems are involved on the equipment. It is completely consistent with my understanding of the qualities of a good troubleshooting electrician, therefore, that I am suggesting in this section that you make it a habit to develop skills and knowledge in many areas outside the immediate field of electrical work.

In Chapter 6 we discussed the process of collecting information. The focus of that chapter was primarily the information pertinent to the particular machine breakdown at hand. There are, however, other levels of information acquisition.

Specific Equipment Information

First, there will be the in-depth information regarding the actual equipment you are working on. If you are in a plant maintenance position, this will include the entire range of the machinery in that plant. Earlier, we used an example of a thermocouple problem that was the result of an electrician improperly wiring the leads. If you have thermocouple applications in your plant, then for you, this level of understanding would include a complete working knowledge of thermocouples and heat controllers. You should know how thermocouples work, the different types available for various heat ranges, the kinds of currents they generate, what the heat controller is sensing, how it is responding to the information, and so on. That kind of information is not a part of your electrical licensing, but it is the information that will give you the unique skills to troubleshoot effectively in a plant where thermocouples are used. Specific thermocouple knowledge may enable you to do a three-minute thermocouple repair by welding a new junction, whereas a fellow electrician may shut the equipment down to replace the entire thermocouple and lead wire system. Since a thermocouple junction is nothing more than a fusion of the two dissimilar leads, you may do the repair with a soft reducing flame from an **oxy-acetylene** torch.

Your in-depth knowledge of the equipment in your plant will include a familiarization with maintenance and operation manuals for all of the equipment in the plant. You will want to learn machine functions well enough so that you can operate basic equipment, though you will include machine operators in your troubleshooting team. You should also become familiar with systems represented on the machines in your plant. Pneumatic and hydraulic equipment are important examples. This is the reason for the closing chapter, *Troubleshooting Hydraulic and Pneumatic Systems*.

Broad Information

Second, there will be the broader information level that will go beyond equipment in your plant. As you develop in this area, you will find, to your amazement, that there are many alternate ways to do things besides the way it is done in your plant, and possibly, besides the way it is done in your particular industry. There are often good reasons why certain conventions are followed in a given plant, and they should not be changed carelessly. But do not always assume that it represents the only or the best way to accomplish the task. Many times a piece of equipment (or control devices on the equipment) was installed the way it was because it represented technology at the time of installation—or possibly it was the result of a low-bid installation. From that time on, all repairs have been done as though each component item represented an optimum standard in the industry, and great care has been taken to replace defective components with exact equivalent replacement parts. During equipment repair, systems can often be upgraded with alternate replacement components for more reliable performance. Upgrading equipment, however, will require specialized and broad knowledge on the part of the one doing the work. The electrician who has taken time to broaden his or her range of knowledge in these areas will bring an employer some valuable assets. These unique knowledge skills become not only the basis of better troubleshooting, but they add significantly to job security and advancement potential.

What has been said of knowledge skills could just as well be said of any number of work skills. The more the electrician understands and is able to do, the greater will be the available number of job opportunities. Equally, that greater skill is going to make a substantial contribution to increased troubleshooting effectiveness.

Each of these three areas (the advantages of trade licensing, the advantage of a small-shop setting, and the advantages of developing higher skill levels) is ultimately concerned with increasing your ability in troubleshooting. Each of them represents an area in which you can develop skills beyond the basic skills of the general electrician. These skills can dramatically improve your troubleshooting speed by increasing your awareness of the equipment you maintain.

TOOLS FOR BROADENING YOUR KNOWLEDGE

Broadening your knowledge—which should make you both a better troubleshooter and a better electrician—will not take place simply because you are working in the trade. It will be the result of a planned learning approach on your part. Each person learns differently. Therefore, you will need to adapt what is being said to your own needs. Nonetheless, be realistic. If your study methods are not producing substantial growth in knowledge, then you need to reevaluate what you are doing. Ultimately, however, if you do achieve the competency in your field that this broader knowledge can bring to you, it will take place because you were willing to do the hard work to get there.

The following areas of effort are suggested as being the most productive in helping you reach your goal:

Reading

Some form of consistent reading is mandatory if you wish to acquire significant knowledge in the trades. In addition to specific technical information dealing with your immediate electrical trade concerns, broad reading in many other areas will also be helpful. Your best ideas will often be based on information from entirely different application sources. Furthermore, an understanding of other subjects may give you insight into the current problems you are dealing with. For example, understanding the process used in the manufacture of solid-state devices may have little immediate application in your plant. Yet your background in that area may prevent future equipment downtime by prompting you to use a better **heat sink** or a snubber when installing a solid-state device.

Wisely selecting your reading materials will be as important as the time spent in reading. Books and magazines will always be important sources of reading material. In addition, do not overlook what is available through equipment suppliers in their technical bulletins and catalogs. There is a genius to the way this material is condensed and organized. Four or five pages in the technical section of an equipment catalog may give the entire contents of an engineering text in summary form. For a brief acquaintance with a wide range of subjects, technical catalogs and bulletins are an invaluable source of information. While you have the catalog in hand, study the main sections as well. Familiarization with the equipment available on the market may help you later when you are upgrading or altering electrical equipment.

Work Experience

Do not overlook the learning opportunities you have in your regular work. This is particularly true when you purposefully attempt work in new areas.

Your attitude, however, is an important part of whether or not this adds to your knowledge. Some electricians can spend their time going through the motions and gain little by the experience. On the other hand, if you are open to information that is available to you through that experience, it can become an important part of your learning activity. Learning to ask appropriate questions can be an important source of information. We are continually around other people in the trades who know things we do not. Fellow tradespersons and equipment suppliers can give you an amazing amount of new information.

You may use specific needs in your plant to acquire information. For instance, many plants with PLC-controlled equipment have purchased the machine from an outside manufacturer. If your plant does not currently design and build equipment using PLCs, you may use the need for a specialized piece of equipment as an opportunity to advance your personal knowledge. You may be able to work with a PLC supplier who will do the basic programming as a part of the sale. You may then enroll in an evening PLC training program. As the project progresses, you can load the program on your own computer and familiarize yourself with it. As the new equipment is put into operation, you can make necessary changes on the timers, counters, or auxiliary circuits to get it running properly. Now your foot is in the door to continue upgrading other equipment in your plant with better PLC controls. Understandably, these opportunities will be greatest in a smaller plant. Larger plants already have specialists.

Seminars and Continuing Education

These sources of information must be carefully evaluated. Some are far too expensive for the value of the information given. Some are just plain worthless. Nonetheless, there are many things available that are well worth the expense and time. Many excellent seminars are provided at no cost by equipment suppliers. These seminars may often be the basis of applicable information for your particular plant maintenance needs. You can frequently question a competent seminar leader who would charge a substantial fee if hired as a consultant.

You will also want to investigate adult continuing education programs. There are an increasing number of trades classes being offered through local community colleges. As these programs develop, more specialized types of information are being offered. In addition to the *National Electrical Code®* classes, you can often find excellent courses in motor controls, solid-state devices, plant maintenance, PLCs, etc. Most colleges with trades classes also offer other related areas of interest to electricians, such as refrigeration, heating systems, and the like. Many of these classes are offered in the evenings.

Equipment Examination

In Chapter 6 you were encouraged to examine equipment after it fails. Take that a step further, trying to think through both the principles of operation of that piece of equipment or component, and then looking for the reasons why it failed from the standpoint of electrical and mechanical theory. Why, for instance, did a motor burn out because the voltage dropped? The coils evidence the same charring as if the motor had run on too high a voltage. The answer is not in merely looking at the motor. It is in understanding the effects of heat produced by overcurrent, which can be a product of either a high- or a low-voltage condition—depending, of course, on the motor's load.

There is an answer to the above question. Electrical power is derived from two components (volts × amperes or watts). If a motor is supplied with a lower-than-normal voltage, and if the power output (horsepower) can proportionately drop (as in the case of a blade fan), the current could remain approximately the same. In this case, the motor would not be in jeopardy. On the other hand, if the voltage dropped but the motor power output remained constant, as, for example, with a loaded conveyor belt, the motor amperage would need to rise proportionately to maintain the same horsepower. If the current rises, the motor coils will be subject to additional heat. So too, if the voltage rises (as in the case of the VOM coil mentioned in Chapter 5), the current will be proportional to the voltage. Thus, either a high or a low voltage could burn out a motor if the load was constant in the low-voltage condition.

Advanced Licensing or Certification

An excellent way to increase your understanding of electrical problems is to continually work toward advanced licensing. Plan on qualifying for your master electrician license and taking the exams. In some states, these licenses are called supervisor or administrator licenses. Not only will advanced licenses enhance your job opportunities, but they will be a significant means of upgrading your electrical knowledge.

There are also numerous areas in the trades where further certification can be sought. Pursue these areas if they are consistent with the trades area in which you are employed. This may include such areas as boiler operator or welding certification.

Becoming a Plant Foreman or General Manager

In large manufacturing plants, most state electrical licensing requirements favor the electrician with a supervisor (or administrator) level license as a maintenance shop foreman. That same licensing advantage may move the electrical supervisor (or administrator) to the plant general manager's job.

You would be wise to consider your interest and ability for these future management positions. The person who eventually works into a management position is not "better" than one who spends an entire career in the applied trade. However, if you want to move in the direction of management or some other form of further responsibility as you get closer to retirement, you should begin planning toward that goal now. Future management opportunities will undoubtedly be conditioned by more than your skill as an electrician. You will certainly need management skills. These skills may be evidenced by your ability to work with fellow employees. In larger plants, however, the human resource office's evaluation of your management skills may be based on your willingness to enroll in evening business classes. In most cases, when opportunities are given to a tradesperson to move into management, it will be based on that individual's broad range of technical ability and desire to learn.

You will find that as you get older it becomes harder to obtain advanced licensing. Apply for higher licenses as soon as you qualify. Do not wait until you need it. Not only might a downturn in the economy demand the licensing if you want to be among those still working, but having your advanced license as early in your career as possible will allow you to move on to other supplementary skill acquisition.

A CONCLUDING THOUGHT

Why has so much attention been given to the subject of learning in a troubleshooting text? It has been done for a very basic reason. Troubleshooting is, by nature, an attempt at finding a yet unknown malfunction. The malfunction may not even be electrical. In the final analysis, there are really only two approaches to isolating the problem. The least effective method is to start at the beginning of the circuit and check everything as it comes until the electrical malfunction is found. The efficient approach is to determine the most likely area of malfunction and confine testing to that precise area.

How does an electrician determine the most likely area of malfunction? Obviously, that determination is dependent on the electrician's understanding of the equipment being tested. The more completely the equipment is understood—and the more information the electrician possesses about the total range of systems involved—the more effective the troubleshooting process will be.

The purpose of this book is to teach you an effective troubleshooting technique. On-line troubleshooting can save you a great deal of downtime. However, it is not the rote use of the technique that will produce results. Effective results will only be achieved by the electrician who has an appropriate background of knowledge in conjunction with the skills to use the troubleshooting technique.

There is much that you can do in the electrical trades beyond troubleshooting. Greater professional effectiveness and responsibility, however, will demand broadening of your skills and knowledge.

CHAPTER REVIEW

Effective troubleshooting is measured by three important components: (1) the speed of the troubleshooting process, measured as the total machinery downtime; (2) the reliability of the diagnosis and repair, measured by the absence of future downtime caused by undiagnosed or poorly repaired malfunctions; (3) cost effectiveness, measured by the ratio of the least expensive means of putting the equipment back into service, while at the same time maintaining the lowest cost from lost production.

CHAPTER 12 Broadening the Electrician's Horizons

Since effective troubleshooting is more than a problem-solving technique, it follows that the electrician will need to apply broad knowledge and judgment in order to accomplish the task. A background of technical information will greatly contribute to the effectiveness of the troubleshooting process. Though field experience is an important part of the electrician's learning process, it is by no means the only area of concentration. An appropriate study of practical and theoretical written material should be a part of the electrician's plan for skill improvement.

The work context can make an important contribution to the electrician's overall knowledge. Of the many work-related ways in which broader background knowledge can be gained, three were mentioned: (1) Electrical trade licensing gives the electrician the unique opportunity of developing multiple trade skills without violating Code and safety requirements. (2) Employment in a small shop may give the electrician opportunity to gain background in numerous mechanical and hydraulic areas. (3) The electrician should be consciously aware of the advantage of developing unique skills in any employment or avocation situation.

The electrician should have a planned learning approach to the information available in the trade fields. Acquisition of new information will come in many forms. Reading, work experience, seminars and continuing education, equipment examination, and advanced licensing or certification were mentioned as areas worthy of attention.

The need for greater knowledge and learning has been emphasized in this electrical troubleshooting text because, by its very nature, troubleshooting is an attempt at finding a malfunction outside the normal electrical circuit function. The skill of the troubleshooting electrician will be highly dependent on his or her ability to appropriately use information to isolate unknown fault areas.

THINKING THROUGH THE TEXT

1. What are the three components of effective electrical maintenance? Can you give a brief description of each?

2. What advantage is gained in electrical troubleshooting through possessing greater background knowledge?

3. What advantage does the licensed electrician gain over other plant maintenance personnel?

4. According to the viewpoint of this chapter, what specific benefits are to be found in reading?

5. Can you explain how the examination of failed equipment is a part of the electrician's learning experience?

6. In what way can advanced licensing contribute to more effective electrical troubleshooting?

7. Troubleshooting is, by its very nature, an attempt at finding an unknown variable in the electrical circuit. How does greater knowledge on the part of the troubleshooting electrician aid in diagnostic work?

CHAPTER 13

Troubleshooting Hydraulic and Pneumatic Systems

OBJECTIVES

After completing this chapter, you should be able to:
- Read a basic hydraulic or pneumatic print.
- Understand the basic function of hydraulic or pneumatic valve controls and actuators.
- Know how to use a hydraulic or pneumatic diagram to locate electrical functions for troubleshooting.
- Understand the basic maintenance precautions while working on hydraulic or pneumatic systems.

Why should an electrician study hydraulics or pneumatics? The answer will, of course, depend on your specific job. Nonetheless, if you are working in any kind of manufacturing plant, there is a high probability that you will be servicing electrical systems that are controlling hydraulic or pneumatic equipment. Your value as both a general maintenance electrician and a troubleshooter will be greatly enhanced if you understand basic hydraulic or pneumatic systems.

How *involved you* eventually become with hydraulic maintenance is a decision you alone will make. There is a great demand in industry for the electrician who can cover a wide range of maintenance tasks. The purpose of this chapter, however, is not to give a course in hydraulic or pneumatic repair. It is merely to give enough information so that you can knowledgeably troubleshoot the electrical system on a hydraulic- or pneumatic-actuated piece of equipment.

If you are interested in further study of hydraulics, we would suggest the *Industrial Hydraulics Manual,* published by Vickers.[*] This book is an excellent source of information with numerous examples. Although very complete, the book is easy to follow whether you are just beginning in hydraulics or are quite familiar with the subject. Similar materials are available for pneumatic systems.

For the purposes of this book, most of what I need to say about either hydraulic or pneumatic electrical systems is similar. Consequently, with the exception of introducing the basic pneumatic symbols,

---
[*] Vickers, Inc., Rochester Hills, Michigan. The book is available through Vickers' hydraulic dealerships.

184 CHAPTER 13 Troubleshooting Hydraulic and Pneumatic Systems

I will emphasize hydraulic systems. You can then apply this information to electrical troubleshooting on pneumatic systems.

HYDRAULIC DIAGRAMS

Before studying the specific hydraulic symbols, it may help you to look at the partial hydraulic print diagram in Figure 13–1. The hydraulic diagram is a pictorial representation of the fluid circuit. In this diagram you see three actuators (the components that convert hydraulic pressure into work) at the top of the diagram. The mold space adjustment actuator is a hydraulic motor, and the clamp cylinder and ejector cylinder are both double-acting cylinders. These cylinders can apply mechanical pressure in both extension and retraction. You also see the valves that control the hydraulic flow. On the left-hand side of the diagram under the mold space adjustment motor, you see valve number 44, which is used to control the fluid flow to that motor. Each valve shows the fluid flow for any one of two or three valve positions. Valve 44 has three positions; valve 41 directly under it has only two positions. Finally, you see lines connecting the valves, actuators, and other parts of the system.

There are four symbolic categories in a hydraulic circuit diagram.

Pumps

You will always see a pump system on a hydraulic circuit. Conventionally, it is at the bottom of the drawing.

Lines

The diagram will show the rigid supply lines and hoses as lines on the drawing. The lines are not la-

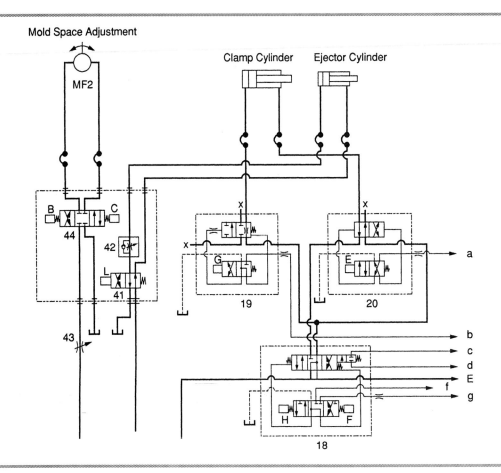

Figure 13-1   A portion of the hydraulic circuit for a plastic injection molding machine.

beled with numbers or letters as they are in electrical drawings. However, their physical location in relationship to the components they are serving identifies their function.

Valves

All hydraulic systems are controlled by valves. There is a wide range of valve functions, from controlling set pressures to determining the direction of flow. Valves may be manually, electrically, mechanically, or pressure (pilot) operated. You should also be aware that there are two important valve functions: **direct operation** and **pilot operation**. In some cases where the hydraulic volume is small, a direct-acting valve arrangement can be used. Valve 41 shown in Figure 13–1 is direct acting. This electrically operated valve is used to control the actual hydraulic fluid circuit to the cylinder.

However, in most cases, the hydraulic fluid volume requires a larger valve than practically could be controlled by an electric solenoid. Figure 13–1 also shows an example of a larger flow valve controlled by a pilot valve. Valve assembly 20 consists of two valves; the electrical solenoid-operated pilot valve is used to control a small-volume hydraulic circuit. The small-volume hydraulic circuit in turn controls the main valve. The main valve is thus *hydraulically*, not electrically, operated.

Actuators

Generally, the hydraulic fluid is acting against either a hydraulic motor or a cylinder. The hydraulic motor produces rotary motion, whereas the hydraulic cylinder produces linear motion. All hydraulic systems can be visualized as producing one or the other motion. Motors and cylinders are called actuators.

If you are not yet familiar with hydraulic diagrams, you should have no difficulty quickly learning how to read them. Once you understand the meaning of each of the symbols, you will then merely trace the circuit to understand its function.

However, there is one area of explanation necessary in order to avoid confusion. You will later see a component enclosure symbol drawn as a box with a broken line. The symbol merely indicates the location or physical grouping of the valve(s) or component(s). It means that the valves are enclosed or are a part of an assembly, so that they form a unit rather than individual components. The important distinction is that the lines do not represent hydraulic circuits. In Figure 13–1 you will see a broken-line box drawn around valves 41, 42, and 44. These lines indicate that the three components are enclosed inside of the equipment or are mounted as a single unit. Component enclosure lines are not hydraulic circuit lines.

HYDRAULIC SYMBOLS

Table 13–1 shows the most commonly used hydraulic symbols. These symbols conform to the specifications of the American National Standards Institute (ANSI).

Because this is only an overview of hydraulics, individual symbols are not explained. The following brief explanations, however, deal with some of the less obvious symbol meanings.

Line Functions

1. *Working lines* are the main lines carrying fluid to or from the pumps and actuators. They are represented as heavy dark lines.

2. *Pilot lines* are the control lines that carry the hydraulic fluid from a control (pilot) valve to a high-volume main valve. Generally, these are passages drilled in the mounting blocks; they are not pipes or hoses. They are represented as light or long broken lines.

3. *Drain lines* are used to carry nonworking (low-pressure) hydraulic fluid back to the tank. They are represented as short broken lines.

4. *Flexible lines* are usually high-pressure, steel-reinforced, rubber-jacketed hose. This is in contrast to rigid lines, which are usually steel pipe.

5. *Restricted lines* are either fixed or adjustable. They are used to limit the travel speed of a cylinder or motor by limiting the flow, but not the pressure, of the hydraulic supply line. In actuality, this is a hydraulic component rather than a physical line, although it is represented as a line function.

186 CHAPTER 13 Troubleshooting Hydraulic and Pneumatic Systems

Table 13-1 Standard hydraulic symbols. *Adapted from* Industrial Hydraulics Manual. *Courtesy of Vickers, Inc.*

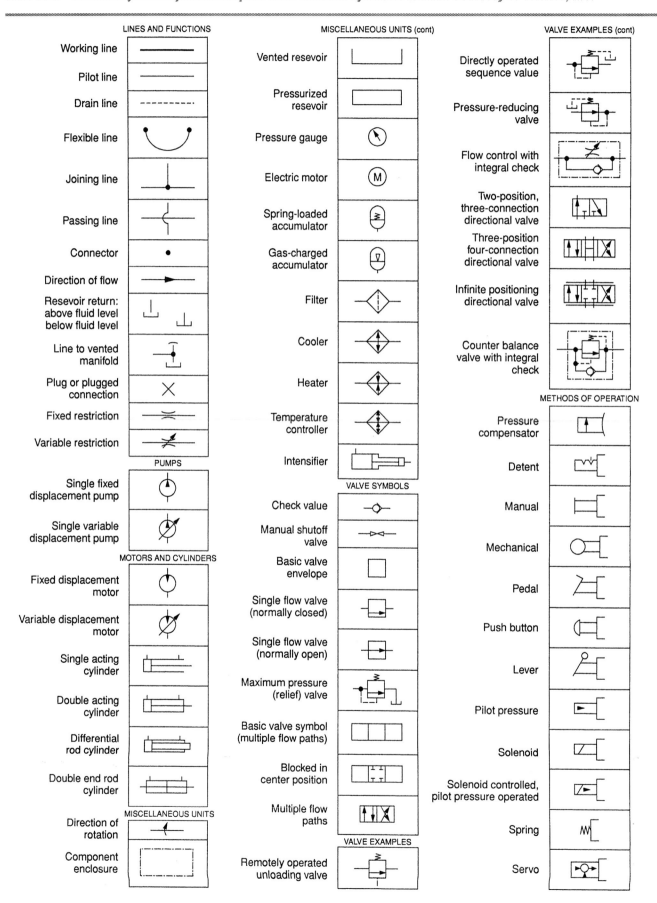

Pumps

Hydraulic pumps supply hydraulic fluid to the actuators in the system. Pumps are either fixed- or variable-displacement units. A **fixed-displacement pump** has a given output volume for a given motor speed. The displacement of a variable pump can be controlled to give varying output volumes depending on need even though the motor is running at a constant speed. The **variable displacement pump** is used as an energy-saving device. A directional arrow, which indicates the inlet and outlet sides of the pump, may or may not be used in the hydraulic schematic.

Actuators

Actuators are the mechanical components that convert hydraulic flow to mechanical motion. They are either motors (rotary motion) or cylinders (linear motion). Though there are a number of different cylinder configurations, the symbols should be self-explanatory.

Valves

Valves represent the most complex part of the diagram. Since the valve can be drawn to represent a number of different functions, each valve drawing must be studied individually to determine its precise action.

Figure 13–2 shows a four-way, two-position valve. The figure shows a possible application in a control circuit. On the left-hand side of the figure, the solenoid is shown as being de-energized. A pump is shown supplying pressure through the valve to the retraction side of the cylinder. In this application, the cylinder will always remain retracted under full working pressure as long as the solenoid remains de-energized. The right-hand side of the figure shows the solenoid as being energized. This causes the valve to shift to the left and reverses the pressure flow, causing the cylinder to extend under full pressure. Notice that the hydraulic fluid on the rod side of the cylinder can vent through the valve to the tank.

Figure 13–3 shows a four-way, three-position valve. This valve uses two solenoids and two springs to position the valve in any one of the three positions. You can see that when either solenoid is energized, the valve will shift to the right or left, resulting in the same control function as the valve in Figure 13–2. However, if neither solenoid is energized, the spring will center the valve and the ports will be blocked. In the center position, the hydraulic pressure will be removed from the actuator. Valve ports are identified as they are in this figure. A and B ports connect with the actuator. The P (pressure) port is connected to the pump side of the hydraulic system. The T (tank) port is used for fluid return and is vented to the tank.

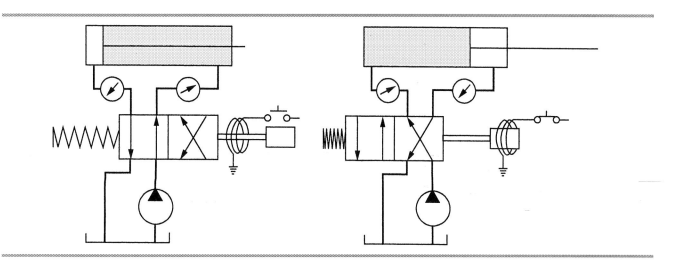

Figure 13-2 A far away, tow-positions value. The drawing on the left shows a de-energized Solenoid that retracts the cylinder. The drawing on the right show an energized solenoid that extends the cylinder and allows the hydraulic fluid on the rod side of the cylinder to vent through the value to the tank.

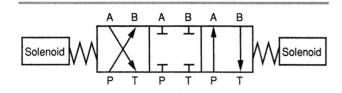

Figure 13-3 A four-way, three-position valve. The valve used two solenoids and two springs to position the valve in any one of the three positions. A and B port connect with the actuator. The P (pressure) port is connected to the pump side of the hydraulic system. The T (tank) port is used for fluid return to the tank

There are many valve configurations available. These two examples do not show all of the fluid paths that can be created with hydraulic valves.

1. *Check valves* are used as one-way valves; the fluid will flow in one direction but cannot return from the high-pressure side of the system to the low-pressure side. Check valves are often paired with restricting valves to give full flow in one direction and controlled flow in the reverse direction.

2. *Valve envelopes* are used to indicate the function of the valve in a given position. A directional valve is used to direct the flow of the hydraulic fluid. A typical directional valve will have either two or three positions. However, it is identified as either a two-way valve or a four-way valve because of the number of flow paths. The valve is always drawn in its center—or rest—position. The **valve envelopes** are the flow paths for that valve for each of its two or three positions.

    The valve is drawn so that you see the flow path when the valve is in its normal position. In many cases, that position will be blocked so that there is no flow. The symbol for a blocked valve port is a capped line drawn like a T. To visualize the flow path in other positions, you need to mentally move the envelope to the other position(s).

    In order to avoid confusion, you must remember that only one envelope is operational at any given time. The number of envelopes for a given valve shows the number of working positions in which that valve is capable of operating. To understand each of the various flow paths for that valve, you must mentally shift the valve to its other envelope positions.

3. *Valve operation* is designated with a symbol outside of the envelope. Electrical solenoid operation has already been mentioned. Other symbols would indicate that the control valve is manually operated, has spring centering or return, or has some other specialized form of operation.

Most hydraulic diagrams number the components such as valves and pumps. You will then find a legend section on the diagram that identifies each of the components by the diagram number. Typically, the information that is given includes the valve function, the valve replacement numbers, and the pressure settings or other pertinent operating information. Study the legend section of the hydraulic diagram in Appendix B.

TROUBLESHOOTING WITH A HYDRAULIC DIAGRAM

The hydraulic diagram can give useful information for general troubleshooting as well as specific electrical troubleshooting. A brief example of how that might work in both cases is shown.

In Chapter 7, you were working with the ejector system on the ladder diagram (solenoid I on line 31). Now see how the hydraulic diagram could help you. In both of the following examples, the ejector system is not functioning.

As was the case in the example in Chapter 7, you are again told that there is a problem with one of the molding machines in the production area. The ejection system is not functioning. In this case, however, it will not function in *any* mode. If you remember, in the previous example the ejection system would work in the manual setting but not in the automatic mode.

Again, you make a preliminary visual equipment and safety check and find nothing wrong. This time, however, the machine has not been in production. The condition is different in that the setup person from the last shift started changing the die, but left the job to be completed by your shift's setup crew. The machine is ready to operate, with the exception of the ejection cycle, which does not function.

Using the Hydraulic Diagram to Locate Electrical Components

In the previous examples, you located the circuit diagram lines you needed to work on from the electrical diagram itself. In this case, however, you have a copy of the hydraulic diagram with you. Therefore, as a first step in your electrical work, you start with the hydraulic diagram. Figure 13–4 shows the hydraulic circuit for the ejector cylinder. Since the cylinder is not extending, you know that the line that is not pressurizing is the one entering the closed end of the cylinder. That is, of the two lines supplying the ejector cylinder, the line you will want to check is the one drawn on the left of the cylinder symbol.

You trace the line through the entire circuit and discover only two control areas between the cylinder and the pump. The first is valve 41, which is controlled by solenoid L (ejector). The second is control valve 42.

Since you are doing your electrical testing first, you now know that you need to verify the operation of solenoid L. As you recall from the procedure in the previous example, there are two solenoids on which the operation of the ejector cylinder is dependent. The first solenoid is L, which you have just identified. The second solenoid is M—ejector pressure. Solenoid M is at the lower right-hand corner of the diagram. If you are adept at reading hydraulic diagrams, you will see how this controls the line pressure by shifting to pressure-relief valve 36. For the present example, however, it may be easier for you to locate the solenoids in the circuit from the electrical diagram.

In the same way as you have done before, you have the operator run the machine until it stalls on the ejection cycle. Since you have identified solenoid L (ejector) from the hydraulic diagram as being the checkpoint, you locate that solenoid on the ladder diagram and proceed with a voltage check on the wire leading to the solenoid.

If your first voltage test indicates an electrical problem, you will continue working with the electrical system. However, for the sake of this example, you find that the electrical system is fully operational. Now you will need to check the hydraulic system itself.

Using the Hydraulic Diagram for Component Checking

When you trace the lines from the ejector cylinder to the pump, you notice that there are two control areas. The first control is valve 42, which is labeled as the throttle check valve. This valve is a manual speed-control valve.

Logic says that something could be wrong with valve 42. When you check the valve, you discover that, for some reason, it has been turned to its closed position. The setup person from the last shift apparently made the adjustment and said nothing to the next shift. It is not an adjustment that is frequently changed and it was overlooked as a possible cause of the ejector malfunction. Opening the valve and recycling the machine will indicate that the problem is solved.

Depending on what the actual cause of the problem had been, you could have used the hydraulic diagram to help you locate an area for electrical testing or to locate specific hydraulic components that need testing. Certainly not all hydraulic testing will be this simple. Yet, you would be surprised at how many times both hydraulic and electrical problems are simply and quickly solved if you can use diagram information to *isolate the probable problem areas.*

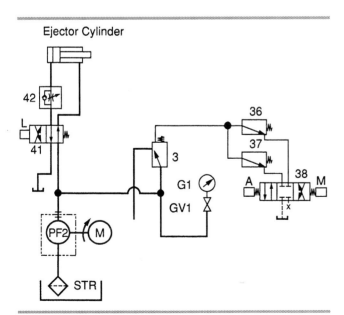

Figure 13-4 The left-hand side of the drawing shows the hydraulic pump and circuit for an ejector cylinder. The right-hand side of the drawing shows the pressure regulation circuit for the ejector.

ELECTRICAL PROBLEMS IN THE HYDRAULIC SYSTEM

Your initial task in troubleshooting either a hydraulic or pneumatic system is to determine if the problem is an electrical control problem or a hydraulic/pneumatic problem. As a rule of thumb, if the hydraulic or pneumatic circuit completely fails to function, it is an electrical problem. On the other hand, erratic, slow, or noisy operation—or the presence of excessive heat or leaks—is an indication that the problem is in the hydraulic/pneumatic part of the system. By far the higher percentage of problems will be in the electrical controls.

In earlier chapters, the subject of troubleshooting control components has been adequately discussed. The one remaining area that you will occasionally need to work with is the solenoid coil. In Chapter 9, you were told how to use a clamp-on ammeter for coil testing. That test procedure will supply valuable information when you work with coils. It will allow you to determine if the solenoid is failing to mechanically shift, which would indicate a mechanical problem with the valve itself. If the current draw is high, indicating that the spool valve is not shifting, the valve will require mechanical repair. Refer to Figure 13–5 for an example of this type of testing in an electrical panel.

Generally, a problem with the solenoid coil will require either a simple lead wire repair or replacing the coil. You will occasionally find loose connections on the coil terminals. This is especially true if wire nuts are used and the terminals have been reconnected a number of times. After tightening the connections, another check should be done to verify the system. If, however, a continuity check of the coil shows any fault, the coil should be replaced. For the most part, coils will give long service, but when they evidence shorting or deterioration, there is nothing that can be done to salvage them; they should be replaced. Spares should be kept in stock.

TROUBLESHOOTING AND REPAIRING HYDRAULIC SYSTEMS

The information given in this overview section is generally true for any hydraulic system. However, when troubleshooting or repairing a specific hydraulic system, information from the equipment supplier may be needed.

Figure 13-5 This is an actual test of a hydraulic solenoid valve done from the electrical panel. The solenoid is being checked by measuring the current on wire 50. This testing procedure saves considerable time over physically dismantling the valve.

System Design

When working on hydraulic systems, it is best to assume that the original system as designed by the manufacturer was properly engineered for its intended service. At times, field modifications will be required because of the addition of new functions or because replacement parts for that specific machine are no longer available. However, a seemingly uncomplicated procedure such as relocating a subsystem or changing a component part can cause unexpected problems. Because of the need to keep all repairs and modifications within the engineering parameters of the original design, the following suggestions should be kept in mind:

1. Each component part in the hydraulic circuit must be compatible with the entire integrated system. For example, if you are replacing or adding an inlet filter on the pump, it must be ade-

quately sized to handle the flow rate of the system. Failure to match the filter to the flow rate could cause **cavitation**, hammering from air in the fluid, and subsequent damage to the pump.

2. All lines must be properly sized and be free of restrictive bends. Undersized or restricted lines may result in decreased equipment performance because of reduced flow rates.

3. Some components must be mounted in specific positions relative to other parts of the system. Modifications may jeopardize the performance of the hydraulic system. For example, the housing of a **flooded suction** in-line pump must be lower than the fluid level in the hydraulic reservoir. The pump housing must remain full of hydraulic fluid in order to provide proper lubrication when the pump is started.

4. The addition of adequate test points for pressure readings, while not essential for operation, will reduce troubleshooting time. Permanently mounted gauges or quick-disconnect fittings for portable gauges allow faster evaluation of the system's performance.

Familiarity with the System

Knowledge of the operation of the specific hydraulic system you will be troubleshooting is probably the greatest aid to fast and effective diagnostic work. Every component in the system has a purpose. The construction and operating characteristics of each should be understood. For example, knowing that a solenoid-controlled directional valve can be manually actuated may save considerable time in isolating a defective solenoid.

Any assembled hydraulic or pneumatic equipment supplied by a reputable manufacturer will include complete documentation for both the electrical and hydraulic systems. A plant electrician working on this machinery would do well to thoroughly read both the electrical and hydraulic/pneumatic sections of the documentation. Make it a point to understand *both* sections before attempting troubleshooting or repair work even if your part of the job is confined to the electrical controls.

In addition to the general operation of any given hydraulic system, the following specific items are areas of information you should have available to you when you are troubleshooting a faulty hydraulic system:

1. You should know the capabilities of the specific system you are working on. Each component in the system has a maximum rated speed, torque, or pressure. Loading the system beyond the specified limits greatly increases the possibility of system failure and hazard.

2. You should know the correct operating pressures. The overall hydraulic system will have a maximum set pressure, as will individual subsystems. For example, the pressure-relief valve for the entire system may be set at 2,200 psi, while an individual hydraulic motor in the system may have a maximum operating pressure of only 1,500 psi and will be controlled by its own pressure-relief valve. Always set and check pressures with gauges. Do not rely on guesswork for high-limit pressure settings.

    A complete hydraulic schematic should include the set pressures for each pressure-relief valve in the system. In cases where specific pressure settings are not specified on the drawings, the following procedure may be used to set secondary relief valves.

    Be certain that the main pressure-relief valve(s) are set at the specified pressure limit. Under no circumstances should a secondary relief valve be set at a higher pressure than the main system relief valves. Reduce the secondary relief valve to its lowest pressure and energize the hydraulic system. The system will generally stall. Increase the set pressure until the system operates normally. The correct operating pressure is the lowest pressure that will allow adequate performance of the system. It must, however, be a pressure below the maximum ratings of the machine and within the operating ranges of the hydraulic components.

3. You should know the proper signal levels if you are servicing servo-control systems. **Servo valves** are infinitely positioned valves. Unlike a conventional spool valve, which controls only the direction of the fluid, the servo valve can control the amount as well as the direction of the

fluid. Thus, the valve can be used to control the position, speed, or acceleration of an actuator if the appropriate feedback sensing devices are included in the electrical circuit. Troubleshooting servo-control systems will include hydraulic and electronic troubleshooting skills. You should also be aware that servo systems are much less tolerant of hydraulic oil contamination than conventional pilot-operated spool-valve systems.

Develop Systematic Procedures

As is the case in any troubleshooting work, you need to begin by carefully observing the system. Component failures may often be identified by either abnormalities evident in the component itself, or by a characteristic action of the hydraulic system. If not caused by low control voltages, a humming or noisy solenoid valve indicates a jammed or nonfunctioning spool. A creeping hydraulic cylinder indicates worn piston packing or possibly leaking control or check valves. Develop a logical sequence for setting valves, mechanical stops, interlocks, and electrical controls. Start in the areas that have the greatest influence on the entire system. If there is a pressure problem in a cylinder, check the main pressure-relief valves if the line pressure is low, rather than starting with the assumption that the failure is in the cylinder itself. Tracing of flow paths can often be accomplished by listening for flow in the lines or feeling them for warmth.

Recognize Indications of Failure

The ability to quickly identify component failures in a specific system is usually acquired with experience. However, the following indicators are usually a sign of impending failure:

Excessive Heat

Excessive heat indicates abnormal conditions. Heat may be the result of electrical, hydraulic, mechanical, or heat-exchanger problems. Hydraulic problems often generate heat. For example, a warmer-than-normal tank return line on a relief valve indicates continuous operation at the relief valve setting pressure. Hydraulic fluids that have a low **viscosity** will increase the internal leakage of components, resulting in a heat rise. Cavitation and slippage in a pump will also generate heat. Mechanical problems may generate heat. A misaligned motor coupling places an excessive load on the bearings and can be readily identified by the heat generated.

Excessive Noise

Excessive noise indicates wear, misalignment, or cavitation. Contaminated fluid can cause a relief valve to stick and chatter. Pump noises (usually caused by cavitation) can be caused by dirty filters or fluid, high fluid viscosity (because it is cold or the wrong grade), excessive drive speed, low reservoir level, loose intake lines (which allow air into the fluid), or actual mechanical problems such as a worn pump or coupling.

Practice Appropriate Maintenance

Four simple maintenance procedures have the greatest effect on hydraulic system performance, efficiency, and life. Yet, the very simplicity of them may be the reason they are so often overlooked.

1. Maintain clean hydraulic fluid of the proper type and viscosity in the reservoir. Keep the level within the normal operating range.
2. Change filters and clean strainers at appropriate intervals.
3. Keep all connections tight and leak-free, eliminating air inclusion in the hydraulic fluid.
4. Periodically sample the hydraulic fluid for outside laboratory analysis for metal, particulate (dirt), and moisture contamination.

PNEUMATIC SYMBOLS

Table 13–2 shows the basic pneumatic symbols. Many of these symbols are readily identifiable if you understand hydraulic symbols. In general, lines, valve envelopes, pilot- versus direct-operated valves, and the like are similarly represented.

Unique Features to Pneumatic Diagrams

The foremost difference between hydraulic and pneumatic diagrams is the absence of a pump in the

Table 13-2 Standard pneumatic symbols.

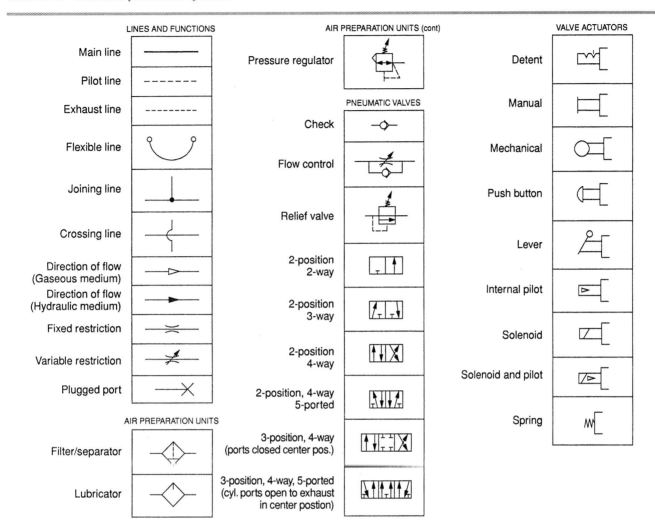

pneumatic diagram. Unless the machine is a self-contained unit, the air source is external to the diagram.

Hydraulic systems frequently have individual pressure regulation for sections of the hydraulic system. That is, individual cylinders may have their own pressure regulation. Though there may be individual pressure controls on pneumatic equipment, it is less frequently encountered. Flow controls, however, *are* frequently used in hydraulic equipment.

In general, hydraulic flow control is always before the actuator, whereas in pneumatic systems the control is after the actuator. A typical pneumatic cylinder speed control will be mounted on the exhaust side of the system.

Another difference between the two is the direction of the fluid after it has been used in the actuator. A hydraulic system must return the oil to the tank. Because the oil becomes heated in the pumps and valves, adequate means of cooling and filtering the oil are required. On the other hand, pneumatic systems generally vent the spent air to atmosphere either directly at the point of use or at the exhaust port of the valve. Consequently, hydraulic systems always require return lines. A pneumatic system must use a return line if the actuator is double acting. A single-acting cylinder or motor may require only one line. Since the air is not operating in a closed system, cooling is not required beyond the air compressor itself.

Work Safely

Safe working practices are mandatory during either hydraulic or pneumatic maintenance and troubleshooting, just as they are in electrical troubleshooting. *In addition to the precautions you take* against electrical hazard, you will also need to be

194 CHAPTER 13 Troubleshooting Hydraulic and Pneumatic Systems

aware of the unique safety requirements of the hydraulic equipment. Again, as stressed in Chapter 1, good judgment is probably the most important factor in overall hydraulic safety. Keep your hands away from moving or potentially moving equipment. Do not manually override hydraulic pilot valves unless you have set the machine so that it can operate safely. Be aware that most hydraulic fluids are flammable. Above all, take every necessary precaution in locking out both electrical and hydraulic or pneumatic valves. Where applicable, use lockable valve covers so that someone else cannot open a valve when you are working on the system. Finally, block out all necessary moving cylinders or hydraulically activated machinery that may represent a hazard.

A DAY AT THE PLANT

Never adjust a pressure-relief valve to a higher setting with the motor off. Set the pressure with the system in operation and watch a pressure gauge while you make the adjustments. This prevents the system from surpassing its upper limits. This caution comes as a result of an unforgettable experience. I was rebuilding the hydraulic system on a large hydraulically operated machine that had a maximum working pressure of 2,200 psi. I had replaced a number of valves, including the main pressure-relief valve. I mounted a quick-disconnect fitting and gauge on the valve to set the pressures. When the installation was complete, I started the motor from the panel and then walked to the back of the machine to look at the gauge. In the time it took me to realize what I was seeing, the pointer moved from 2,900 to over 3,000 psi and was still climbing. I set a speed record between the back of the machine and the big red button on the operator's panel! That was the last relief valve I ever installed without first backing the pressure screw to the lowest setting.

Pneumatic systems have an inherent danger in the compressibility of the air. Unlike hydraulic fluid, which cannot expand, air can store—and release—dangerous amounts of energy after it is shut off at safety valves. *Always lock out and then release residual air pressure before working on pneumatically operated equipment.*

Aside from the hazards inherent with moving machinery, probably the greatest danger area related to hydraulic equipment is the presence of fluid under high pressure. The system used for the examples in this chapter is set for 2,000 psi on the main lines. Look at the legend section of the diagram in Appendix B. Valves 1, 3, and 5 are set at 2,000 psi working pressure. At these pressures, bursting lines or fittings can be lethal. Never put your hand in front of a leak—the fluid can penetrate the skin and cause severe damage.

CHAPTER REVIEW

A basic understanding of hydraulic or pneumatic systems may greatly enhance your value as a maintenance electrician. If nothing else, the knowledge will help you work more effectively when you are troubleshooting electrical controls on the equipment.

The hydraulic or pneumatic diagram is a pictorial representation of the system. It will include symbols for pumps, lines, valves, and actuators, which are generally motors or cylinders. A typical diagram will identify valves (usually with numbers) with a legend on the side. However, supply lines (or hoses) have no specific identification markings. It is necessary to visually trace each line to determine its function.

Hydraulic or pneumatic flow control valves are used to control the direction of the oil or air. A specialized symbol called an envelope is used to designate each of the flow possibilities for a given valve. The envelope is shown on the drawing in its normal (non-actuated) position. To determine the other flow patterns, the adjacent envelopes are mentally moved into their operational position(s). Flow control valves are generally either two-way valves (valves with two positions and two flow paths) or four-way valves (valves with three positions and four flow paths).

Hydraulic or pneumatic diagrams may be used effectively as a troubleshooting aid in either of two ways. They can be used to locate electrical components by tracing hydraulic circuits through their respective flow control valves. The diagrams can also be used to trace components in a given circuit.

Special mention was made of the symbol for the component enclosure. This symbol, which is usually a box drawn with a broken line, surrounds either a valve assembly or a group of valves and components in a common location. The component enclosure symbol indicates the location or physical grouping of the valves and components and is not indicating hydraulic lines or circuits.

When you are doing the initial troubleshooting on an inoperative hydraulic or pneumatic system, you will first need to determine if the malfunction is mechanical or is in the electrical controls.

When working with actual hydraulic equipment, it is important that you view the entire hydraulic design as an integrated system. Any modifications or repairs must be compatible with the system. In order to maintain the system effectively, you must understand how it works and be aware of the various system pressures.

Understanding and practicing good safety procedures is mandatory in both hydraulic and pneumatic work. You must be aware of the high pressures and the hazards encountered because of them.

THINKING THROUGH THE TEXT

1. Draw and identify the hydraulic symbols for a pump, a two-way directional valve, a check valve, a blocked valve port, and a double-acting cylinder.

2. What is the difference between a direct- and a pilot-operated flow control valve? Why is the pilot-operated valve used?

3. Draw a component enclosure symbol. What does this symbol mean?

4. Draw the symbol and describe the purpose of each of the following lines: a working line, a pilot line, and a drain line.

5. What are the typical main line hydraulic pressures (rated in psi) mentioned in this chapter?

6. List two indications of hydraulic component failure. Give examples of each type of failure.

7. What specific hazards are mentioned as areas demanding attention when working on either hydraulic or pneumatic systems?

APPENDIX

Complete Electrical Diagram

Electrical Diagram Legend

Selector Switches

S1	Operation
S2	Screw operation
S3	Mold open-close
S4	Plunger forward-return
S5	Screw unit forward-return
S6	Mold height adjust
S7	Mold height

Toggle Switches

SW1	Zone 1 heater power
SW2	Zone 2 heater power
SW3	Zone 3 (nozzle) heater power
SW5	Control power
SW8	Press slow
SW9	Fast mold-close
SW10	Intrusion
SW12	Screw speed slow

Limit Switches

LS1	Mold open
LS2	Mold lock-up
LS3	Single cycle
LS5	Feed
LS6	Screw unit stroke
LS7	Slow down
LS8	Ejector forward
LS9	Ejector stroke
LS10	Rear gate safety
LS11	High pressure
LS12	Low pressure
LS14	Emergency mold-open
LS15	Ejector return
LS20	Lubrication

Push Button Switches

E-STOP	Emergency stop
PBM	Motor start
PBS	Motor stop
PBE	Ejector
PBL	Lubrication light

The control diagram was supplied by Kawaguchi, Inc., Buffalo Grove, Illinois. It is the electrical diagram of the Kawaguchi JEKS 180 plastic injection molding machine. Wire numbers and symbols were altered to accommodate American electrical schematics.

198 APPENDIX A   Complete Electrical Diagram

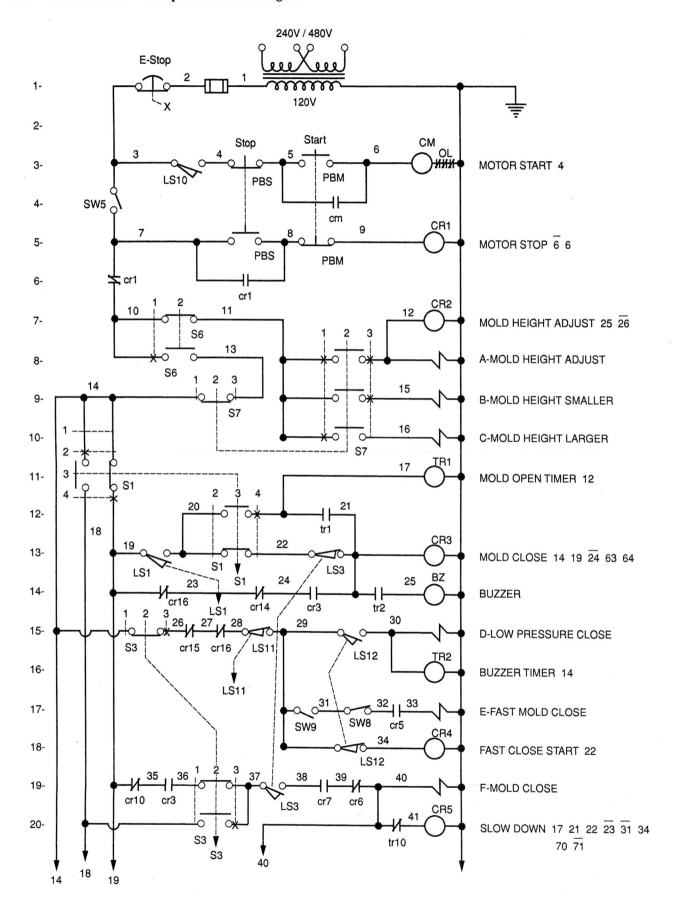

APPENDIX A Complete Electrical Diagram

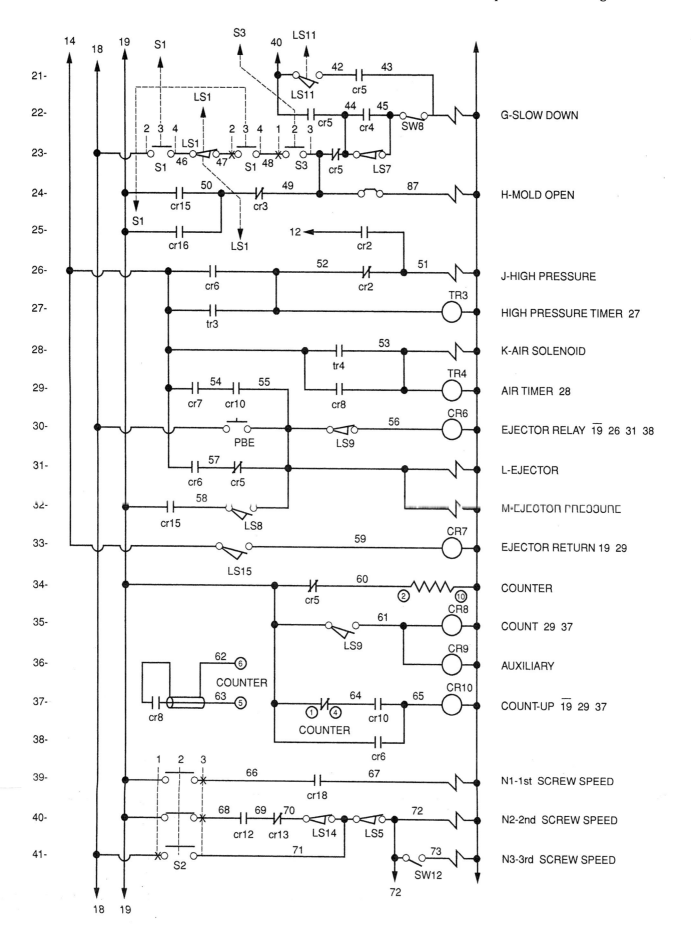

200 APPENDIX A Complete Electrical Diagram

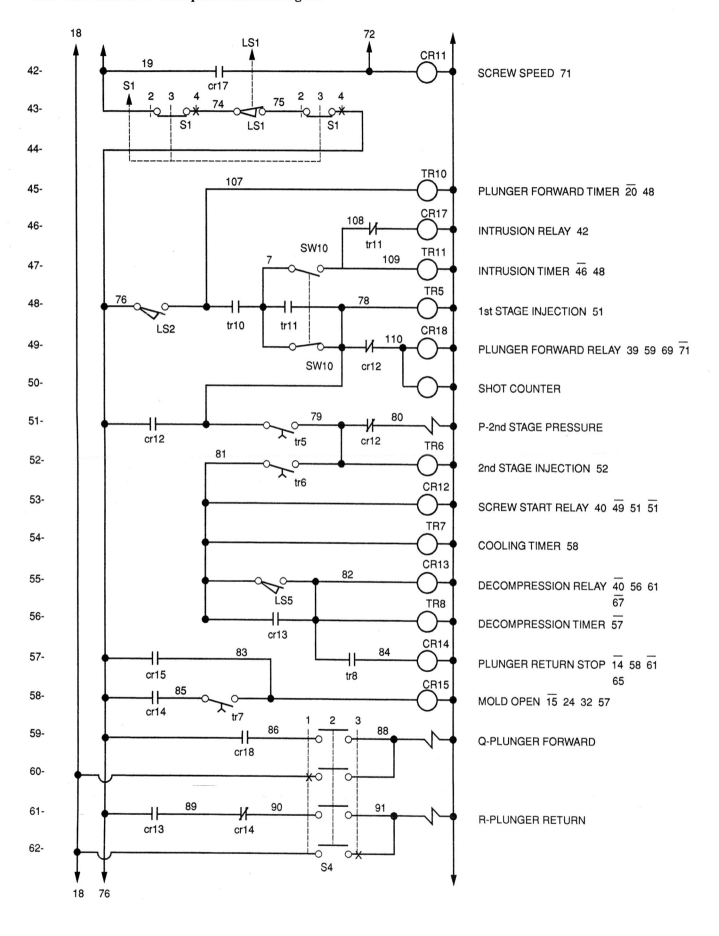

APPENDIX A Complete Electrical Diagram

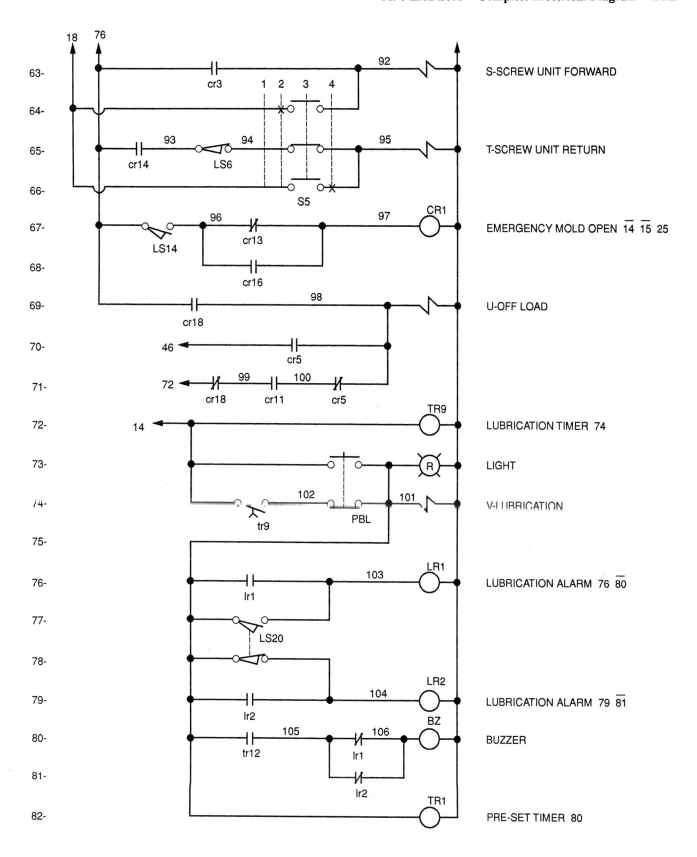

APPENDIX A Complete Electrical Diagram

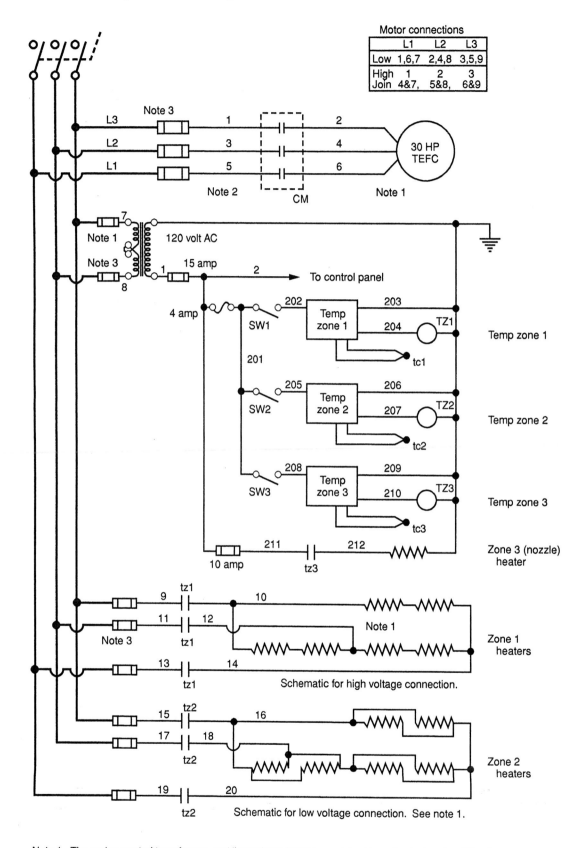

Note 1 - The motor, control transformer, and line voltage heaters are wired from the factory for 480 volts.

Note 2 - All line leads (240 or 480 volt) are black. All control leads (120 volt) are red. The control neutral is white.

Note 3 - Size motor, control transformer primary, and line voltage heater fuses according to information supplied for high (480 volt) or low (240 volt) connections.

APPENDIX

B

Complete Hydroulic Diagram

Hydraulic Components

M	Motor	22KW × 6P	
PF1	Pump	21G SQP21-21-8-1BB-10	
PF2	Pump	8G	
MF1	Hyd. Motor	MIC-033	
MF2	Hyd. Motor	H-170	
HE	Heat Exchanger	FK-1604B	
STR	Suction Strainer	PF-120	
G1	Pressure Gauge	BU-3/8 (3500 PSI)	Main Pressure
G2	Pressure Gauge	AU-3/8 (3500 PSI)	Back Pressure
G3	Pressure Gauge	BU-3/8 (3500 PSI)	Hyd. Motor
GV1	Gauge Valve	GV-03-02	G1
GV2	Gauge Valve	GV-03-02	G2
GV3	Gauge Valve	GV-03-02	G3
1	Relief Valve	BG-03-3233	Set at 2000 PSI
3	Relief Valve	BG-03-3233	Set at 2000 PSI
5	Relief Valve	BG-03-3233	Set at 2000 PSI
7	Solenoid Valve	DSG-03-3C0-RQ115-4022	SOL-J—Rccpo Line Procoural SOL-N3—No. 2 Screw Speed
8	Sol. Pilot Valve	DSHG-06-3C5-E-R115	SOL-N1—No. 1 Screw Speed; SOL-N2—No. 2 Screw Speed
13	Angle Check Valve	CRG-06-5-3013	Minimum Line Pressure
15	Reducing Valve	RG-03-B-2016	Pilot Pressure for Spool Shift
16	Pilot Valve	SD-859-20	Shuts Off Flow to Valve No. 18 when Safety Gate Is Open
18	Sol. Pilot Valve	DSHG-06-3C5-E-RG115	SOL-F—Mold Close; SOL-H—Mold Open
19	Sol. Pilot Valve	DSHG-06-280-E-R2-RG115	SOL-G—When De-Energized Throttles Flow to Clamp Cylr
20	Sol. Pilot Valve	DSHG-06-280-E-R115	SOL-E—Faster Mold Close
21	Flow Control Valve	FG-01-8-N-1103	Speed of Carriage Cylinder
22	Check Valve	SD-862-11	Keeps Pressure in Carriage Cylr
23	Solenoid Valve	DSG-01-3C40-A115-3123	Carriage Cylinder FWD/RTN
24	Sol. Pilot Valve	DSHG-06-3C40-E-RG115	Plunger FWD/RTN
25	Throttle Check Valve	SRC-06-4-4111	Controls Speed of Injection
26	Check Valve		Stops Injection Cylinder Flow
28	Relief Valve	BG-03-3221	Back Pressure Adjustment
29	Flow Control Valve	FG-03-125-2110	Screw Speed
31	Relief Valve	DG-01-22	1st Stage Injection Pressure
32	Relief Valve	DG-01-22	2nd Stage Injection Pressure
33	Relief Valve	DG-01-22	Controls Lock Up Pressure Max at 16000 PSI
34	Relief Valve	DG-01-22	Low Pressure Closing
35	Solenoid Valve	DSG-01-3C2-A115-31	SOL-P—2nd Stage Pressure; SOL-D—Low Pressure Closing
36	Relief Valve	DG-01-22	Ejector Pressure
37	Relief Valve	DG-01-2216 (1100 PSI)	Pressure on Hyd. Motor
38	Solenoid Valve	DSG-01-3C2-A115-31	SOL-M—Ejector Pressure; SOL-A—Hyd. Motor Pressure
39	Relief Valve	DG-01-2216 (1100 PSI)	
40	Solenoid Valve	DSG-01-2B2-A115-31	SOL-U—High Pressure on P1 Pump
41	Solenoid Valve	DSG-03-2B3-RQ115-4025	SOL-L—Ejector FWD
42	Throttle Ck Valve	SRCG-06-H-4111	Ejector Speed
43	Needle Valve	GCTR-02-3107	Mold Height Adjust Speed
44	Solenoid Valve	DSG-01-3C2-A115-31	Hyd. Motor Turning Direction
60	Throttle Valve	SNS-385	

This diagram was supplied by Kawaguchi, Inc., Buffalo Grove, Illinois. It is the hydraulic diagram of the Kawaguchi JEKS 180 plastic injection molding machine.

204 APPENDIX B Complete Hydraulic Diagram

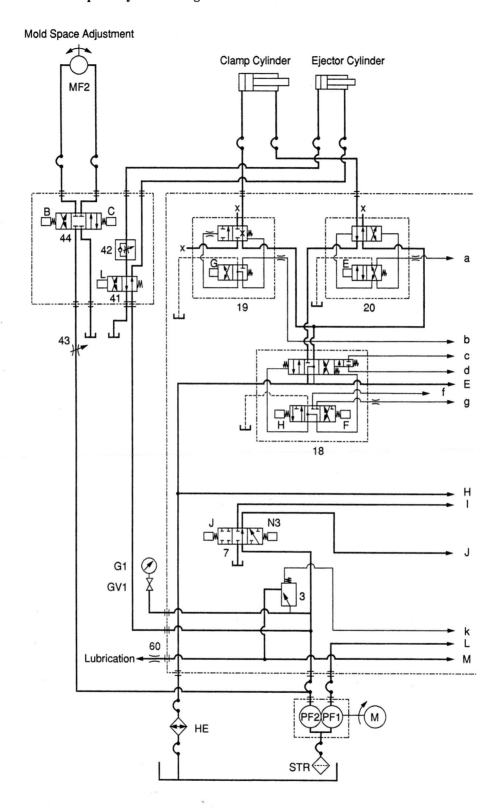

APPENDIX B Complete Hydraulic Diagram

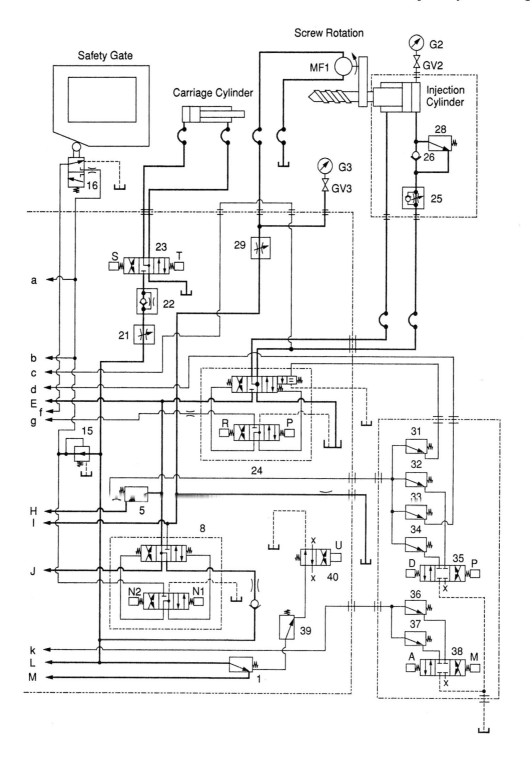

Key Terms

Actuator, electrical device A cam, arm, or similar mechanical or magnetic device used to trip limit switches.

Actuator, hydraulic or pneumatic A cylinder, motor, or other component that does work.

Alphanumeric A character set that contains both numerical and alphabetic characters.

Alternating current Current changing both in magnitude and direction.

Ampere (A) Unit of electrical current.

Analog device Apparatus that measures or displays continuous information (e.g., current or voltage). The analog signal has an infinite number of possible values.

Analog meter See **Meter, moving-coil.**

Apparent power (VA) The unit of electrical pressure or potential.

Armature The moving ferrous element in an electromechanical device such as the rotating part of a generator or motor or the moving part of a relay or solenoid.

Automatic mode An operation mode in which all machine operations are independent of operator control.

Backfeed A circuit path through components that are series-connected to the measurement being performed but are not the normal working current path.

Branch circuit That portion of a wiring system that extends beyond the final overcurrent device protecting the circuit.

Breakdown voltage The voltage at which a disruptive discharge takes place, either through or over the surface of insulation.

Capacitance, capacitor Two conductors separated by an insulator.

Capacitor, electrolytic A capacitor consisting of two conducting electrodes, with the anode having a metal oxide film formed on it.

Cavitation The effect of high velocity gas on the surface of a pump vane from gas entrapment in the fluid.

Circuit-breaker An automatic device that opens under abnormal current in the carrying circuit. A circuit-breaker is not damaged on current interruption. The device is ampere, volt, and horsepower rated.

Clamp-on ammeter See **Meter, clamp-on**

Common In reference to conductors, having electrical contact with insignificant interposing resistance.

Component A general term used to mean electrical equipment. In this text, the term *power component* is used to identify the equipment that utilizes current, such as a solenoid or contactor.

Conductor A substance that easily carries an electrical current.

Connection, dynamic A low-resistance connection between conductors that permits movement at the interface.

Connection, static A low-resistance connection between conductors rigidly affixed at the interface.

Contact block, switch The electrical control portion of an industrial-rated switch device. The contact block is generally a removable component that allows configuring the switch function to the circuit.

Contact bounce The continuing making and breaking of a contact after the initial engaging or disengaging of the contact.

Contact testing See **Testing, contact**

Contact, electrical 1. Continuity; the absence of significant resistance between two conductors. 2. The conducting members of a relay, switch, or connector that are engaged or disengaged to open or close an electrical circuit.

Contactor, motor A device for repeatedly establishing and interrupting an electrical power circuit.

Continuity A conductive path for an electrical current from one point in an electrical circuit to another.

Control circuit The circuit of a control apparatus or system that carries the electrical signals directing the performance of the controller but does not carry the main power circuit.

Control circuit voltage The voltage provided for the operation of shunt coil magnetic devices.

Control panel An enclosure in which electrical control devices are mounted.

Controller See **Contactor, motor**

Cross-training Skill development in areas outside of the direct trade of employment.

Current See **Ampere**

Current-carrying An energized conductor.

Current-carrying capacity The maximum amount of current that a conductor can carry without heating beyond a predetermined safe limit.

Current limiting fuse See **Fuse, current limiting**

Cycle An alternating current waveform that begins from a zero reference point. It reaches a maximum value and then returns to zero. It then reaches a negative maximum value and returns to the zero reference point.

Device A unit of an electrical system that is intended to carry but not utilize electrical energy.

Device terminal The electrical connection on a device; frequently a screw terminal.

Diagram A schematic or logical drawing showing an electrical or fluid (hydraulic or pneumatic) circuit or logical arrangements within a component.

Dielectric The insulating material between metallic elements of any electrical or electronic component.

Digital The representation of numerical values by means of discrete numbers.

Diode A solid-state device having two electrodes, a cathode and an anode. Current passes at low resistance in only one direction.

Direct current A continuous nonvarying current in one direction.

Direct operation A hydraulic valve that uses no pilot device.

Disconnect, motor A switch used in a motor branch circuit. It is rated in horsepower and is capable of interrupting the maximum operating overload current of a motor of the same rating at the same rated voltage.

Disconnecting means A device that allows the current-carrying conductors of a circuit to be disconnected from their source of supply.

Documentation An orderly collection of diagram information covering the control and line voltage systems of an installation. These records provide valuable reference data for installation and maintenance of the equipment.

Downtime The time when a system is not available for production due to required maintenance either scheduled or unscheduled.

Effective voltage The total of all voltage values imposed on a conductor.

Electrical equipment The electromagnetic, electronic, and static apparatus as well as the more common electrical devices.

Electrical system An organized arrangement of all electrical and electromechanical components and devices in a way that will properly control the machine or industrial equipment.

Electrolyte A substance in which the conduction of electricity is accompanied by chemical action.

Electromagnetic field The field of influence produced around a conductor by the current flowing through it.

Electromagnetic interference (EMI) Electromagnetic phenomena, which, either directly or indirectly, contribute to degradation in performance of an electronic system.

Electromechanical A term applied to any device in which electrical energy is used to magnetically cause mechanical movement.

Exposed As applied to electrically live parts, capable of being inadvertently touched or approached at less than a safe distance by a person.

Fault An accidental condition in which a current path becomes available that bypasses the connected load.

Feeder The circuit conductors between the service equipment and the branch circuit overcurrent device.

Flooded suction A hydraulic pump inlet located under fluid level. The pump is self-priming.

Frequency Number of complete variations made by an alternating current per second; expressed in hertz. See **Hertz**

Fuse An overcurrent protective device with a fusible member that is heated directly by the current passing through it. The fusible member is melted with excessive current to open the series circuit.

Fuse, current-limiting A fuse that limits both the magnitude and duration of current flow under short-circuit conditions. By definition, the current-carrying member will melt within one-half cycle.

Fuse, dual element Often confused with time delay. Dual element is a manufacturer's term describing fuse element construction.

Fuse element A calibrated conductor that melts when subjected to excessive current. The element is enclosed by the fuse body and may be surrounded by an arc-quenching medium such as silica sand. The element is sometimes referred to as a *link*.

Fuse, one-time A fuse with a fixed current limit, irrespective of time.

Fuse, time delay A fuse with combined thermal elements providing continuous load protection at a fixed current limit, yet capable of sustaining considerably higher current loads for short intervals of time. These fuses protect a motor during high inrush starting currents. The current/time requirements are defined in the UL 198 fuse standards.

Fused disconnect Generally an air-break switch in conjunction with a fuse.

Grounded Connected to earth or to a conducting body that serves in place of the earth.

Grounded conductor A conductor that carries no current under normal conditions. It serves to connect exposed metal surfaces to an earth ground to prevent hazards in case of breakdown between current-carrying parts and exposed surfaces. If insulated, the conductor is colored green, with or without a yellow stripe.

Grounding jumpers Wire connections intentionally placed between a circuit conductor and ground.

Hard-wire Permanently wired; the conductors are firmly attached to devices and loads at terminal connectors.

Harmonic A sinusoidal wave having a frequency that is an integral multiple of the fundamental frequency. For example, a wave with twice the frequency of the fundamental is called the second harmonic.

Heat sink A mounting base, usually metallic, that dissipates or radiates into the surrounding atmosphere the heat generated within a semiconductor device.

Hertz (hz) International unit of frequency equal to one cycle per second of alternating current.

High resolution See **Meter, high resolution**

Holding (latching) circuit The circuit that maintains a holding output device, such as a relay, after it has been initially energized. The output device is deenergized when the holding circuit is opened. See **Initiate circuit**

Holding (latching) relay A relay that is self-maintaining.

Hydraulic system As used in this text, consists of a high-pressure fluid system for the transfer of energy between a fluid pump and the end use. Most commonly encountered hydraulic systems use either petroleum-based oils or nonflammable glycol-based fluids; 2,000 pounds per square inch of pressure is common.

Impedance (Z) Measure of opposition to flow of current, particularly alternating current. It is a vector quantity and is the vector sum of resistance and reactance. The unit of measurement is the ohm.

Induction, induced field The act or process of producing voltage by the relative motion of a magnetic field across a conductor.

Infinity A hypothetical amount larger than any assignable amount; in resistance measurements, it generally indicates an open circuit.

Initiate circuit The circuit that initially energizes a holding output device, such as a relay, at which time a second circuit within the output device itself holds the output device in an energized state. See also **Holding circuit**

Input device Any connected equipment that will supply information to the central processing unit, such as switches, push buttons, sensors, or peripheral devices. Each type of input device has a unique interface to the processor.

Input port The terminal of a PLC that receives information.

Inrush current In inductive equipment (such as motors, solenoids, or coils), the steady-state current taken from the line with the armature blocked in the rated maximum open position.

Instantly operating contact An output device contact, such as a relay contact, that has no time-related function. It changes state instantaneously in response to the circuit that controls it.

International Electro-technical Commission (IEC) The international standards body that has established safety criteria for hand-held test instruments.

Interposing device An output device that isolates currents or voltages between the control source and the output. Typically, an interposing relay is used between a PLC and the high-current relays or motor starters the PLC is controlling. An interposing relay is a pilot device.

Interrupting capacity The highest current at rated voltage that a device can interrupt.

Inverter See **Variable frequency drive**

Ionize, ionized air The detachment of electrons from outer layers of air molecules at high energy levels. The air is *ionized*. Ionized air has greater conductivity and thus more readily allows arcing. At the highest energy levels, it becomes plasma.

Jog, jogging, inching A quickly repeated closure of the circuit to start a motor from rest to accomplish small movements of the driven machine. A jogging circuit is always momentary.

Jumper A short length of conductor used to make a connection between terminals or around a break in a circuit.

Kcmil The designation of wire dimension given in thousands of circular mils. The *k* is the metric symbol for 1,000, the *c* represents its circular dimension, and the *mil* is 1/1,000 of an inch. Kcmil now replaces the former designation, MCM.

Ladder diagram An industry standard for representing relay-logic control systems.

Ladder element Any one of the elements that can be used in a ladder diagram, including relays, switches, timers, counters, etc.

Latching circuit See **Holding circuit**

Latching relay See **Holding relay**

Light-emitting diode (LED) A diode that emits light when biased in the forward direction.

Limit switch A switch operated by some part or motion of a power-driven machine or equipment to alter the associated electric circuit.

Line numbers The consecutive vertical numbers on the left-hand side of an electrical diagram. Line numbers identify the ladder rungs, not electrical wires.

Line voltage The voltage level of the main branch circuit to the equipment.

Load terminal The lug or connecting means associated with high current conductors.

Lockout A mechanical device that may be set to prevent the operation of a switching device.

Low-resolution meter See **Meter, low-resolution**

Maintained In reference to a device such as a selector switch or push button, it is identified as *maintained* when it will remain in the selected position without hand pressure or mechanical force.

Manual operation A control setting or operation mode that requires selection and starting of each sequence by a machine operator.

Meg- Prefix abbreviation used for a million.

Megger A trade name of AVO International/Biddle Instruments for their megohmmeters.

Megohm (MΩ) One million ohms.

Megohmmeter See **Meter, megohmmeter**

Metal oxide varistor (MOV) A semiconductor device in which the resistance drops rapidly beyond a predetermined voltage value. It is used as a protective device because it can block voltages higher than its rated value.

Meter, analog See *Meter, moving-coil*

Meter, auto selection A test meter capability to select function automatically based on the input from the probes after they are in contact with an energized circuit. The selection is usually between the values of alternating voltage, direct voltage, resistance in ohm value, and continuity. Very few meters have this capability; not all meters have the same auto selection range. See also **Meter, function**

Meter, autoranging A test meter capability to select the intensity of voltage, or the decimal place in resistance, after the probes are in contact with an energized circuit. This is a common feature in electronic digital (DMM) meters.

Meter, clamp-on A meter that uses movable jaws for placement around a conductor. This capability is limited to meters that do not require contact with an energized conductor, such as ammeters and the current portion of wattmeters.

Meter, digital multimeter (DMM) An electronic meter with a digital display capable of multiple function use. Generally, the functions of AC and DC voltage, resistance, and current to 10 amperes are included. Better quality meters will have additional functions.

Meter, function The differing electrical values that a given meter is capable of measuring. Volt AC, Volt DC, OHM, and AMPS are all meter *functions*.

Meter, high resolution A meter that numerically displays true voltage or ohm values.

Meter, low resolution A meter that uses LEDs (light-emitting diodes) and audio signals to indicate predetermined voltage values and continuity.

Meter, megohmmeter A high-range ohmmeter used for measuring insulation resistance values and other high resistances.

Meter, moving-coil An analog meter in which a coil pivots between permanent magnets.

Meter, multimeter A general designation for either a *moving-coil meter* or an electronic meter with a digital display capable of multiple function use. Generally, the functions of AC and DC voltage, resistance, and low current in milliamperes are included.

Meter, ohmmeter A direct-reading instrument for measuring electric resistance. Its scale is usually graduated in ohms, megohms, or both.

Meter, overload protected An overload protected meter generally uses fuses on the input leads. Overcurrent will blow the fuses to protect the meter circuit.

Meter, power factor A direct-reading instrument for measuring power factor. Its scale is graduated directly in power factor.

Meter, protected ohmmeter A meter with special circuitry that protects it from damage when the ohmmeter is momentarily used on a 120-volt control circuit.

Meter, proximity A non-contact meter which is sensitive to the electromagnetic field surrounding a conductor. The meter can sense the presence of a voltage but not the intensity.

Meter, range The maximum voltage or current a meter can read at a given setting. The range selection is a subdivision of function. Autoranging meters do not have required manual range selection. See also **Meter, autoranging**

Meter, resolution The display precision of a meter; the resolution may vary with the range selection. With digital displays, the resolution is determined (for a given range) by the decimal point location and the significant figures to its right.

Meter, volt-ohmmeter (VOM) A moving-coil test instrument with ranges for measuring voltage and resistance.

Microfarad (mfd, mf, or μF) One-millionth of a farad.

Milliampere (mA) One-thousandth (.001) of an ampere.

Millisecond (ms) One-thousandth of a second.

Millivolt (mV) One-thousandth of a volt.

Mode As used in this book, *mode* is the setting for operation of a test instrument or a machine.

Momentary In reference to devices such as selector switches and push buttons, a device is identified as *momentary* when it will not remain in the position selected without continuous hand or mechanical force.

Motor overloads Motor current protection. Motor overloads monitor current as a function of time.

Multimeter See **Meter, multimeter**

National Electrical Code® (NEC®) An electrical installation standard *published by the National*

Fire Protection Association. State agencies adopt and amend *NEC®* for local use.

NEMA National Electrical Manufacturers Association.

Nominal voltage The utilization voltage. See the appropriate NEMA standard for device voltage ratings.

Non-contact testing See **Testing, non-contact**

Nonlinear load Electrical loads in which the instantaneous current is not proportional to the instantaneous voltage; electrical loads in which the load impedance varies with voltage.

Normal position The position of the relay contacts when the coil is not energized, or the position of a spring-loaded mechanical device when at rest.

Normal state The relaxed state of a device; the position of a switch when no force is exerted.

Normally closed (NC) When applied to a magnetically operating switch device such as a contactor or relay or to the contacts thereof, this term signifies the position taken when the operating magnet is de-energized. The term applies only to non-latching devices. The term may also be used for mechanical switching devices that have a spring return such as a limit switch or push button. A NC device is conductive when de-energized or relaxed.

Normally open (NO) When applied to a magnetically operating switch device such as a contactor or relay or to the contacts thereof, this term signifies the position taken when the operating magnet is de-energized. The term applies only to non-latching devices. The term may also be used for mechanical switching devices that have a spring return such as a limit switch or push button. A NO device is non-conductive when de-energized or relaxed.

Occupational Safety and Health Administration (OSHA) The federal agency mandated with workplace safety and health.

Off delay A circuit that retains an output signal some definite time after the input signal is removed.

Ohm (Ω) Unit of electrical resistance.

Ohmmeter See **Meter, ohmmeter**

On delay A circuit that produces an output signal some definite time after an input signal is applied.

Operator, switch The manipulatable portion of an industrial-rated switch device. Selector switches and push buttons are among the most common operators.

Optical isolation interface Electrical separation of two circuits with the use of an optical coupler.

Optical Isolator A device that couples input to output using a light source and detector in the same package. It is used to provide electrical isolation between input circuitry and output circuitry.

Oscilloscope An instrument used to visually show voltage or current waveforms or other electrical phenomena, either repetitive or transient.

Output device A component that uses low power to perform a control function, such as switching, counting, and timing.

Overcurrent Current in an electrical circuit that causes excessive or dangerous temperature in the conductor or conductor insulation.

Overcurrent protective device A device that operates on excessive current and causes and maintains the interruption of power in the circuit.

Overload Operation of equipment in excess of normal full-load rating or conductor in excess of rated ampacity, which, if it were to persist for a sufficient length of time, would cause damage or overheating.

Overload protection Overload protection is the result of a device that operates on excessive current, but not necessarily on short circuit, to cause and maintain the interruption of current flow to the device governed.

Note: Operating overload means a current that is not in excess of six times the rated current for alternating current motors and not in excess of four times the rated current for direct current motors.

Oxy-acetylene The industrial gases of oxygen and acetylene; the torch used with this gas combination.

Panel See **Control panel**

Parallel circuit A circuit in which two or more of the connected components or contact symbols in a ladder diagram are connected to the same set of terminals so that current may flow through all the branches.

Phase rotation The sequence of individual phases in three-phase power that sequences from phase AB to phase BC to phase CA.

Phase-to-ground A measurement or event between a phase conductor and a grounded conductor.

Phase-to-phase A measurement or event between two phase conductors.

Pilot device A first device that directs operation of a second device; usually a low-power device that controls a high-power device. For example, a float switch is a pilot device that responds to liquid levels. See also **Valve, pilot**

Pilot operation A high power device under the immediate control of a low power pilot device. Typically used in hydraulic systems.

Plasma breakdown See **Ionize**

Plug-in Any device to which connections can be completed through pins and matching receptacles. Used of removable miniature relays.

Pneumatic system As used in this text, consists of a high pressure air system for the transfer of energy between an air compressor and the end use. Most common shop air systems operate at 120 pounds per square inch of pressure or less.

Power component The utilization equipment of electrical power in a machine control environment. Solenoids, motor controllers, and power relays are considered power components.

Power factor The ratio between apparent power (volt-amperes) and actual or true power (watts).

Power quality The measure of transients and electrical noise present on power supply lines.

Power supply The unit that supplies the necessary voltage and current to the system circuitry.

Pressure-relief valve See **Valve, pressure relief**

Process switch Process switches are the devices that monitor conditions such as material level, pressure, temperature, or flow and are mounted on the machinery they are monitoring.

Programmable logic controller (PLC) A solid-state machine *control* that uses digital logic. It replaces electromechanical relay panels functioning as control logic. This type of controller is particularly useful in the control of processes, materials handling, and certain machine functions.

Protected ohmmeter See **Meter, protected ohmmeter**

Proximity meter See **Meter, proximity**

Pump, fixed-displacement A hydraulic pump with a predetermined volumetric displacement. Volume output is a function of shaft rotation speed.

Pump, variable displacement A hydraulic pump in which the volumetric displacement can be infinitely altered between zero and maximum output for that pump design.

Push button switch A switch in which a button must be depressed each time the contacts are to be opened or closed.

Reactive power (VAR) The reactive voltage times the current, or the voltage times the reactive current, in an alternating current circuit. Also called wattless power.

Relay A device that is operative by a variation in the conditions of one electric circuit to affect the operation of other devices in the same or another electric circuit.

Relay logic Machine or system controls that use electromechanical relays as the logic processors. The relay is used to identify a condition as *on* (true), or *off* (false).

Resolution See **Meter, resolution**

Root-mean-square (RMS) The value of an alternating current that will produce the same amount of heat in a resistance as the corresponding value of direct current.

RS232 A communication link standard that lists a maximum 5–volt potential. This standard does not require or imply optical isolation.

Rung The horizontal conductor lines on a conventional circuit (relay logic) electrical diagram.

Schematic A diagram of the electrical scheme of a circuit with components represented by graphical symbols.

Sealed The state of an energized relay, contactor, or solenoid after the armature has fully seated and the current has stabilized.

Secured Complete lockout and blocking of all electrical or mechanical equipment that could pose a hazard during maintenance procedures.

Selector switch A multi-position switch that permits one or more conductors to be connected to any of several other conductors.

Semi-automatic An operation mode in which some machine operations are not automatic while selected portions are *automatic*.

Serial addressable output A recent development in machine controls that uses computer controls and a single cable connection between all control points.

Series circuit A circuit in which all the components or contact symbols are connected end-to-end. All must be closed to permit current flow.

Series coil A series coil is connected in series with a load or the full current. Motor controllers and solenoids are drawn as series coils.

Series resistance Two or more resistances joined end-to-end in a circuit so that there is a single current path.

Service factor (motor) The allowable overload permissible by design. It is indicated by a multiplier, which, when applied to a normal horsepower rating, indicates the maximum continuous loading.

Servo valve See **Valve, servo**

Short-circuit An abnormal connection of relatively low resistance between two points of a circuit. The result is a flow of excess (often damaging) current between these points.

Shunt coil A shunt coil is connected with the full voltage applied. Relays and timers are drawn as shunt coils.

Silicon controlled rectifier (SCR) A semiconductor device that functions as an electrically controlled switch for direct current loads.

Sine wave A wave that can be expressed as the sine of a linear function of time, space, or both.

Single-phase A single alternating current or voltage source.

Sinusoidal wave A wave the displacement of which varies as the sine (or cosine) of an angle that is proportional to time, distance, or both.

Smart meter A self-contained digital panel display meter that has programmable input and output functions. The meter is capable of controlling set points in a machine process.

Snubber, RC (resistor/capacitor) A series resistor/capacitor network installed across an alternating current relay or motor controller coil to dampen transients.

Solenoid An electromagnet with an energized coil, approximately cylindrical in form, and an armature whose motion is reciprocating within and along the axis of the coil.

Solid-state Circuitry designed using only integrated circuits, transistors, diodes, etc.

Speed cycle A machine set-up for diagnostic purposes that reduces set time and travel to minimum values. A speed cycle allows testing on the equipment under actual operating conditions without product waste or excessive time intervals between each sequence.

Spool valve See **Valve, spool**

Stalled As used in this book, the state of a machine when the process controls will not sequence to the next function.

Starter An electric controller that accelerates a motor from rest to normal speed.

Steady-state voltage The nominal voltage in an energized circuit that discounts the presence (or possibility) of transient voltages.

Surge suppressor Electronic equipment or devices that protect a circuit from transients. In electrical control panels, RC snubbers, MOV suppressors, and diodes are used across transient-producing induction coils.

Symbol A widely accepted sign, mark, or drawing that represents an electrical device or component thereof.

Tagout An OSHA-mandated hazard prevention procedure that identifies a system lockout with a tag naming the personnel performing the service.

Temperature controller A control device responsive to temperature.

Temperature detector An instrument used to measure the temperature of a body.

Temperature probe The sensing element used for a temperature detector. Any physical property that is dependent on temperature may be employed, such as the differential expansion of two bodies (bimetal), thermo-electromotive force at the junction of two metals (thermocouple), or the change of resistance of a metal (varistor).

Terminal A point of connection in an electrical circuit.

Terminal block An insulating base or slab equipped with one or more terminal connectors for the purpose of making electrical connections.

Termination point The final utilization equipment.

Testing, contact An electrical test performed by causing the internal circuit of the meter to come into electrical contact with an energized circuit.

Testing, non-contact An electrical test performed without causing the internal circuit of the meter to come into electrical contact with an energized circuit.

Thermocouple A device for measuring temperature where two electrical conductors of dissimilar metals are joined at the point of heat application. A voltage difference directly proportional to the temperature is developed across the free ends of the device.

Three-phase Three coupled alternating currents or voltages of a common output source, 120° out of phase with each other.

Time-delay fuse See **Fuse, time-delay**

Timed closed See **On delay**

Timed contact A contact that is not instantaneous.

Timed open See **Off delay**

Toggle switch A lever-operated two-position snap switch used to open or close circuits.

Transformer Converts voltages for use in power transmission and operation of control devices; an electromagnetic device.

Transformer, control A voltage transformer used to supply a voltage suitable for the operation of control devices. Typically, control transformers have a 480/240–volt primary and a 120–volt secondary.

Transformer, isolation A transformer used to isolate one circuit from another.

Transformer, primary The transformer windings that are connected to the supply power.

Transformer, secondary The transformer windings that are connected to the utilization circuit.

Transformer, step-down A transformer used to reduce voltage levels while proportionately sustaining increased current.

Transient *Rapidly changing action* occurring in a circuit during the interval between closing of a switch and settling to steady-state condition, or any other temporary actions occurring after some change in a circuit or its constants.

Trend plot An oscilloscope trace that records a voltage or current level relative to time. The trend plot records the voltage as a *line* rather than as a waveform. See also **Waveform**

Trigger circuit A momentary circuit that initiates an action. See also **Initiate circuit**

VA See **Apparent power**

Valve, direct operating A working line hydraulic valve controlled by a solenoid valve. See also **Valve, pilot**

Valve envelope The symbolic flow path for each hydraulic or pneumatic valve position. A two-position valve has two envelopes; a three-position valve has three envelopes.

Valve, pilot A solenoid-operated hydraulic valve that supplies a limited fluid volume to control a larger working line hydraulic valve.

Valve, pressure-relief A hydraulic or pneumatic valve that determines the maximum allowable fluid pressure.

Valve, servo Infinitely positioned valves. The servo valve can control the amount as well as the direction of the fluid.

Valve, spool A common valve in hydraulic and pneumatic service that uses an axial moving piston with annular grooves shifting within a ported valve body. The alignment of the annular grooves with porting in the valve body determines flow direction.

Variable frequency drive (VFD) A variable speed drive for alternating current motors. Also called an inverter.

Variable speed drive An inclusive term for both alternating current and direct current solid-state motor speed control equipment.

Viscosity The property of a body that, when flow occurs inside, forces a rise in such a direction as to oppose the flow.

Volt (V) The unit of electrical pressure or potential.

Volt amps See **Apparent power**

Volt-ohmmeter (VOM) See **Meter, volt-ohmmeter**

Voltage rating The maximum allowable voltage for equipment or electronic meters.

Watt (W) The unit of electrical power.

Waveform The shape of an electromagnetic wave.

Wire number A number physically attached to equipment wire ends giving each wire a unique identity in the machine control circuit. The wire number is identified in the electrical diagram as the number in close proximity to the line representing that wire.

Wiring diagram A drawing that shows the physical arrangement of electrical equipment and/or components, together with all interconnecting wiring.

Withstand capability The maximum voltage or current allowed in electrical machinery or electrical test equipment before damage occurs.

Index

A

Actuators, 185, 187
Air compression, 194
Alligator clips, 52
Alphanumeric, 103
Alternating current (AC), 87
    effective voltage, 66
    measurement of, 60
American National Standards Institute (ANSI), 185
Ammeter (clamp-on), 55, 56, 60
    amperage reading, 136
    circuit testing, 115, 116, 119–120
    current comparison test, 118–119
    current tracing, 135
    diagnosis and, 174–175
    insulation problems, 120–121, 137
    peak current value, 119, 120, 125
    power quality analyzer, 166
    product information, 116–117, 119, 120
    reading ranges, 116–117
    sensitivity of, 117
    solenoid coil testing, 190
    solid-state, 120
    troubleshooting with, 115–120, 127
Ampere (A), 136
    and current limiting fuses, 169
Analog meters, 124
Analyzer. *See* Power quality analyzer, 166
ANSI. *See* American National Standards Institute
Apparent power (VA), and power quality analyzer, 166
Armature, 58, 116
Automatic mode, 38, 188
Autoranging meters, 10, 63, 99, 103

B

Backfeeding, 101–102

C

Capacitors, 128–129
    construction of, 129
    DC applications, 3, 129–130
    electrolyte, 105–106, 129
    failure of, 129–130
    megohmmeters, 128–130, 137
    multimeters, 105–106, 128, 130
    power factor correction, 128
    product information, 128, 130
    safety, 129
    testing, 56, 105–106, 112, 128–130, 137
CAT classification, 65, 69, 70
Cavitation, 191
Check valve, 188
Circuit breaker, 43
    fuses and, 167–168, 169
Circuit grounding, 38–39
Circuit testing, ammeter, 115, 116, 119–120
Circuits, 39–40, 45, 88
    energized, 3
    de-energized, 6, 10–11, 43, 74, 75, 99
    interrelated, 97
    operating, 3, 50
    ungrounded ("hot" leg), 38–39, 103. *See also* Ground
    tracing of, 134–136
    verification of, 119–120
Common, 5
Compressor head pressure-relief valve, 3
Computers:
    applications of, 110–114
    hazards in use of, 109–110
    laptop, 108
    and multimeter, 108–112
    optical isolation interface, 109–110
Conductors, 5, 13–15, 30
    diagrams, 13–15
    failure of, 168
    identification of, 105, 112
Conduits, as wire locators, 164–165
Connections: dynamic and static, 152
Contact block, switch, 16–18
Contact testing. *See* Testing, contact
Contactor, 5, 150–152
Contacts, 4, 23
    and ladder diagram, 36–38

Continuity, 3
    multimeters and, 104–105, 112
    and on-line troubleshooting, 45–46, 47, 49, 50, 52, 88, 89, 92, 112
    testing for, 54, 56, 60–62, 65, 73, 75, 92, 104–105, 163
Control circuit, on-line, 3, 88
Control circuit voltage, 39–40, 45
Control panel:
    add-on wiring in, 39
    DC current, 130
    individual relays in, 4
    ground, 39
    power lines to, 36
    troubleshooting, 46–49, 50, 70–73, 88–89
    12-volt system, 35–36
Control relay, 26
Control valves, 194
Current, 6, 115–116. *See also* Ampere; Voltage value
Current-carrying capacity, 28
Current comparison test, 118–119
Current tracing, 134–135
Cycle. *See* Machinery

D

Device, definition of, 27
Device terminal, screws for, 6
Diagrams, 1, 30, 171–172. *See also* Electric symbols; Ladder diagram; Wire diagram
    conductors, 13–15
    contacts, 18–19
    creation of, 164–165
    hydraulic system, 184–185, 188–189, 194–195
    numbering systems, 25–26, 29
    pneumatic systems, 192–193
    practical application of, 29–30
    refrigerator compressor motor circuit, 33–35, 161–162
    switches, 15–23
    wire connections, 26
Dielectric, 129
Digital multimeter (DMM). *See also* Multimeter
    and capacitor 128, 130
    on-line troubleshooting, 45–47, 88, 106–108, 110, 112
    product information, 55–56
    use of, 58, 60, 63–66, 75,
Digital voltmeters (DVM), 55–56, 60, 99
Diode, 166
    and DC current, 130
    testing, 106, 112. *See also* Solid state circuitry
Direct current (DC), 3
    capacitors and, 3, 128–130
    failure in panel, 130
    measurement of, 60, 66
    megohmmeter and, 125
    product information, 129

Disconnect, motor, 6
Documentation, 80
Downtime, 1, 51, 110, 154

E

Electric motors, 121–123. *See also* Motors
    burn out, 81, 121, 122–123, 137, 180
    fuses and, 170
    insulation and, 121–123
Electric symbols, 13–15, 18, 30, 161. *See also* Diagrams; Ladder diagrams
    table of, 14
Electrical equipment:
    moving, hazards of, 118
    maintenance of, 123–124
    upgrading of, 179
Electrician, 174–175, 181–182
    advancement (career) as, 177, 181
    cross-training, 175, 176–177
    education of, 174–176, 180
    field experience, 174–175, 179–180, 182
    information for, 179–180
    knowledge for, 175, 176, 178–179
    licensing (certification) of, 176–177, 180–181, 182
    skills of, 178–180, 182
    small shop employment and, 176, 177–178, 182
    technical training, 174–175
Electrolyte, 105–106, 129
Electromagnetic field, 8, 55
Electromagnetic interference (EMI), 24, 73, 117
Electromagnetism, 58
Electromechanical, 4, 142
Emergency switches, 21
Exposed current-carrying terminals, 10

F

Fault, 168. *See also* Ground fault testing
    testing for, 164
Feeder circuit, 104
Ferm's Fast Finder, 169, 170
First component, 2
Flooded suction in-line pump, 191
Fluid viscosity, 192
Fuse, 5, 167, 172
    and circuit breakers, 167–168, 169
    current-limiting, 167, 169
    and ground circuit, 38–39
    location of, 36
    motor loads, 170
    overcurrent conditions and, 167–169
    product information and, 170–171
Fuse element, 171
Fused disconnect, 5

G

Gauges, 191
Ground circuit, 38–39, 43
Ground fault testing, 127. *See also* Fault
   short, 104
   ungrounded ("hot" leg), 38–39, 103
Ground jumper, 10, 36
Grounded conductor, 35–36

H

Hard-wire, circuits, 4
Heat sink, 179
High-voltage installations, 39–40
Holding (latching) circuit, 36, 96. *See also* Initiate circuit
Hydraulic system, 4, 28, 86, 183–184, 194–195
   actuators, 185, 187
   component enclosure box, 185, 186, 195
   design of, 190–191
   diagrams, 184–185, 188–189, 194–195
   electric problem of, 190
   failures of, 192
   fluid viscosity, 192
   Industrial Hydraulics Manual, 183
   maintenance of, 192
   noise of, 192
   and on-line troubleshooting, 44, 49, 183, 188–189, 190–192
   operating pressures within, 191, 194
   and pneumatic systems, 192–193, 194–195
   pumps, 184, 186, 187, 191, 192
   safety, 193–194, 195
   supply lines, 184–186, 194
   valves, 185, 186, 187–188, 191–192, 194
   symbols, 185–186

I

Idle position, 44
Impedance (Z), 65, 68, 69
Induction, 5–6
Industrial Hydraulics Manual, 183
Infinity, 60, 103
Initiate circuit, 92, 94, 96. *See also* Holding circuit
Instantly operating contacts, 23
Insulation:
   ammeter, 120–121, 137
   electric motors, 121–123
   material failure, 122–123, 127, 167–169
   megohmmeter, 121, 122–127
   testing of, 103, 121–129, 137
Integrated circuits, 4
International Electro-technical Commission (IEC), safety standards, 8–9
Interposing device, 27–28
Inverter. *See* Variable frequency drive
Ionize, ionized air, 8

J

Jog function, 21
Jumper, wiring, 5, 50, 52, 62

K

Kcmil, 169

L

Ladder diagram, 4, 13, 30, 32–43, 97
   contact positions, 36–38
   description of, 33–34
   electrical symbols and, 18, 161
   grounding, 38–39
   horizontal lines, 37–38
   line conductors, 39–41
   numbering systems and, 25–27, 29
   on-line troubleshooting, 44, 90–93, 160, 161
   reading of, 35–38, 161
   rungs and, 33
   stalled machinery and, 44–45
   and troubleshooting, 34, 36–37, 44–45, 71, 85–86, 90
   understanding of, 38–40
   vertical lines, 36–37
   wiring diagram compared to, 32–33, 34
Latching circuit. *See* Holding circuit
Light-emitting diode (LED), 63, 69, 72, 75
Lightning strikes, 7
Limit switch, 3, 15–16, 18, 19, 20
   in diagram, 28
   functions of, 19–20, 86
   safety precautions with, 164
Line conductors, 39–41
Line diagram. *See* Ladder diagram
Line numbers, 25, 26, 30. *See also* Wire numbers
Line voltage, 5
Load terminal, 5
Location numbers, 29
Lockout, 5, 6, 7. *See also* Tagout

M

Machinery, 49. *See also* Motor; Physical equipment
   cycles of, 48–49, 52, 94, 95
   examination of, 79–80
   maintenance of, 78, 82, 110, 123–124
   moving-coil meters and, 58
   normal operation position, 49–50
   operators of, 50, 77, 84–86
   safety with, 44, 86
   stalled, 49–50, 71, 85, 90–91, 102–103, 189
   troubleshooting of, 84–96, 98, 167
Magnetic contactor, 2, 81
Magnetic starter coil, 103
Maintenance:
   electrical equipment, 123–124
   hydraulics, 192

machinery, 78, 82, 110, 123–124
   maintenance of, 78, 82, 110, 123–124
Manufacturers's service center, 155–156
Mechanical, troubleshooting, 152–154
Megger, 110, 123, 128. *See also* Megohmmeter
Megohmmeter, 103
   analog, 124
   battery-operated, 128
   capacitors, 128–130, 137
   DC test voltage, 125
   digital, 124
   ground-fault testing, 127
   insulation testing, 121, 122–127
   motor testing, 127
   product information, 124, 126, 127
   safety and, 124–125, 128
   short-circuits, 127–128
Megohms, 103, 123
Metal oxide varistor (MOV), 143
Meter, 56–57, 70. *See also types of meters*
   audio tone and, 47–48, 51–52, 69–70, 72, 74, 130
   auto selection, 63
   choosing, 67–71
   first readings, 152–153
   function selectors, 60, 63
   harmonic, 56
   moving-coil, 54, 55, 56, 57–60
   overload protected, 8–9, 60, 61, 65
   protection rating, 68, 69
   range, 63
   resistance testing, 70, 89, 93
   resolution, 63, 66
   root-mean-square (RMS), 66–66
   safety, 6, 8–11, 46, 60–61
   scales, 66
   sensitivity range, 72, 87
Microfarad (mfd, mf, or μF)
Milliampere (mA), 56
Millivolt (mV), 152
Motor. *See also* Electric motor; Timer motor
   burn out, 81, 121, 122–123, 137, 180
   contactor, 5, 150–152
   control center (MCC), 122
   disconnect, 5, 125
   megohmmeter and, 127
   moisture and, 123
   overloads, 2, 36
   relay points, 150–152
   reverse rotation, 130–131
   three-phase, 130–131, 149
Moving-coil meters and, 58
Multimeter, 8, 63, 65, 99–100, 112–113. *See also* Digital multimeter (DMM)
   capacitor testing, 105–106, 128, 130
   computer interface with, 108–112
   continuity testing, 104–105, 112

   diode testing, 106
   hazards of, 106
   holding in hand, 106
   lead or conductor identification, 105, 112
   off-line testing, 103–104
   on-line testing, 101, 102
   power quality analyzer, 166
   product information, 107–108, 109–110
   resistance testing, 99, 102, 103–105
   selector switch, 99, 100, 112
   short-circuit testing, 103–104, 127–128, 128
   troubleshooting with, 100–102
   voltage testing, 100–102, 115
Multiple Lockout Device, 7

N

National Electrical Code® (*NEC®*), 5, 27, 167, 175, 176
   grounding and, 35
   overcurrent protective devices, 168–169
   size 1 contactor, 116
   voltage and, 9–10
Neon lamp, 131, 136–137
Nomograph, 125
Non-contact testing. *See* Testing, non-contact
Nonlinear load, and plant power quality, 166
Normal position, 16, 18
Normal state (switches), 15
Normally closed (NC), 15
Normally open (NO), 15

O

Occupational Safety and Health Administration (OSHA), 6
   lockout/tagout procedures, 7
Oersted, Hans Christian, 58
Off delay, 23–24
Ohm (Ω), 68, 69
Ohm's law, 58, 107
Ohmmeter, 46, 64, 104–105
   motor failure and, 121, 122
   protected, 64
   as time-saver, 126
Oil level, 3,
   crankcase, 3
On delay, 23–24
On-line troubleshooting, 3, 10, 50–52, 63, 97–98. *See also* Testing; Troubleshooting
   continuity, testing for, 45–46, 47, 49, 50, 52, 88, 89, 92, 112
   control panel, 46–49, 50, 70, 91–93
   downtime and, 1, 51, 110, 154
   example of, 87–95, 98
   hydraulic systems, 44, 49, 183, 188–189, 190–192
   intermittent problems, 51, 166
   ladder diagrams, 44, 90–93, 160, 161
   mechanical functions, 152–154
   meters and, 67–71

oscilloscopes, 149
planning and, 91
pneumatic systems, 49
power quality, 166–167
procedure for, 45–48, 49–50, 96–98
profitable, 11–12
safety, 4–6, 45, 48, 52, 90–91, 97, 98, 162
testing, 5–6, 46–49, 87–90, 162–163
voltage limitations, 9–10, 71–73, 162
Operator switch, 16
Optical isolation interface, 109–110
Oscilloscopes, 56, 107, 158
bench, 140
local power distortion, 142–143
manufacturers's service center, 155–156
performance-over-time monitoring, 140, 156–157, 159
power quality and, 154–155, 158, 166
product information 139–140, 154
safety and, 142
screen, reading of, 141–142
trend plot testing, 147–154, 158
variable frequency drive, 145–146, 147
waveforms and, 139–140, 142–147, 158
Output device, 36
letters and numbers, 26–27, 28, 30
location of, 38–39
and parallel circuit, 105
Overcurrent damage, causes of, 167–168
Overcurrent protective device, 168–169
Overload (OL), 36–37
protection, 36
Oxy-acetylene torch, 178

P

Panel. *See* Control panel
Parallel circuit, 37, 46–47, 105
Parameter measurements, 166
Peak current value, 119–120, 135
Performance-over-time monitoring, 140, 156–157, 159,
Phase meters, 130–131, 137
rotation testing, 131
Phase rotation indicators, 56, 131
Phase-to-ground, 103–104, 112
Phase-to-phase, 103–104, 112
Physical equipment. *See also* Machinery
failures of, 81–82
troubleshooting, 76–82
Physics, laws of, 175
Pillow block bearings, 3
Pilot device, 27–28
Plasma breakdown, 8. *See also* Ionize
Plug-in unit, 28
Pneumatic systems, 4, 194–195. *See also* Hydraulic systems
air compression and, 194
diagrams and symbols, 192–193

and hydraulic systems, 192–193
on-line troubleshooting and, 49
safety, 193–194, 195
Power component, 24, 28, 30, 36
location of, 38–39
Power factor correction, 128
Power lines, to control panel, 36
Power quality, 154–155, 158, 172
evaluation of, 166–167
Power quality analyzer, 166
product information, 166, 167
Pressure gauge, hydraulic, 89
Pressure-relief valve, 2
compressor head, 3
hydraulic system, 191
Pressure switch, 2
Preventive maintenance. *See* Maintenance
Process switch, 16, 18
Product information, 115
ammeters, 116–117
capacitors, 128, 130
current tracers, 134–135
digital multimeters (DMM), 55–56
DC current, 129
fuses, 170–171
megohmmeters, 124, 126, 127
multimeters, 107–108, 109–110
oscilloscopes, 139–140, 154
phase meter, 132
power quality analyzer, 166, 167
testing equipment, 68, 70, 71, 73
wire sorters, 133, 163
Programmable logic controller (PLC), 4, 35, 74–75
Protected ohmmeter. *See* Ohmmeter
Protective gear, 10
Proximity meter (tester), 5, 6, 71, 135
Pumps. *See also* Water Pump
fixed-displacement, 186
hydraulic system, 184, 186, 187, 191, 192
variable displacement, 187
Push button ejection (PBE), 87
Push button switch, diagrams of, 15, 16, 17, 18, 20, 21–22

Q

Quality Assurance, 156

R

Reactive power (VAR), and power quality analyzer, 166
Refrigerator compressor motor circuit, 33–35, 161–162
Relay, 4, 95–96
armature, 119
with lights, 96
Relay coil, 8, 26, 93, 104, 115, 116
Relay contact, 18–19, 26, 28, 51
numbers, 28
Relay logic, 4

Relay points, 150–152
Repair procedures, 48, 80, 179
Resistance load bank, 144
Resistance values:
   temperature compensated, 125, 126
   testing for, 125–126
Resistor/capicitor (RC) snubbers, 143
Resolution. *See* Meter, resolution
Reverse rotation, 130–131
Root-mean-square (RMS), 66–67
RS232, 109
Rung, 33

S

Safety, 4–10, 77, 177
   capacitors, 129
   circuit grounding, 38–39
   current tracing, 136
   hydraulic systems, 193–194, 195
   limit switches, 164
   machinery and, 44, 86
   megohmmeter, 124–125, 128
   meters (in general), 6, 8–11, 46, 60–61
   oscilloscopes, 142
   pneumatic systems, 193–194, 195
   protective gear, 10
   switches, 21–22
   testing equipment, 7–9, 57, 60–61, 106
   three-phase, 9, 130–131, 149
   troubleshooting, 45, 48, 52, 97, 98
Schematic, 32
Secured, 6
Selector switch, 15, 16–18
   in diagrams, 28
   functions of, 22–23
   in multimeters, 99, 100, 112
Semi-automatic switch, 87
Serial addressable output, 4
Series coil, 24
Series resistance, 58
Service factor (motor), 168
Servo valve, 191–192
Shading coil, 116
Short-circuit, 38–39, 103, 119
   direct, 169
   and megohmmeter, 127–128
   and multimeter, 103–104
   phase-to-phase, 169
Shunt coil, 23
Silicon controlled rectifier (SCR), 143–144, 145
Sine wave, 66
Single-phase, safety standards for, 9
Sinusoidal wave, 67
Smart meter, 4
Snubber, RC (resistor/capacitor), 179

Solenoid, 3, 4, 26, 28, 70, 93
   coils, 118, 190
   for hydraulics, 187, 188, 189
   non-functioning, 44, 45, 46, 47, 162–164
Solid-state circuitry, 4, 8, 99, 114, 136. *See also* Diode testing
   devices, 179
   ammeter and, 120
   multimeter and, 106, 112
Speed cycle, 93–95
Spool valve, 119
Stalled machinery, 44–45, 49–50, 71, 85, 90–91
   hydraulic systems and, 189
   on-line testing and, 102–103
   safety and, 44
Starter, motor, 4
Steady-state voltage, 8
Stop function, 36
Surge suppressor, 143
Switches, 20, 23
   circuits and, 17–18, 20–21
   contacts, 51
   diagrams of, 15–23, 28
   electrical function of, 16
   emergency, 21
   hand and foot, 18
   maintained, 16, 21
   momentary, 15, 21, 22
   push-button, 21–22
   selector, 15, 16–18, 22–23, 28, 99, 100, 112
   two-position, 17
   three and four-position, 17–18
Symbols. *See* Electric symbols

T

Tagout, 6
Technical assistance, 155–156
Temperature:
   probe, 111
   and resistance values, 125
Temperature detector, 125
Terminal block, 11
Termination point, 4
Testing, 5–6, 46–49, 97, 113. *See also names of individual meters;* Testing equipment
   for capacitors, 56, 105–106, 112, 128–130, 137
   circuit tracing, 134–136
   contact, 5, 55–56, 67–70, 70–73, 75, 87, 88, 98
   for continuity, 54, 56, 60–62, 65, 73–75, 92, 104–105, 163
   current comparison, 118–119
   diode, 106, 112
   electric motors, 121–123
   for ground faults, 104, 127, 164
   for insulation problems, 103, 121–129, 137
   motor, 127

non-contact, 5, 55, 56, 70–71, 72–73, 75, 87, 88
remote control, 109
for resistance, 99, 102, 103–105, 125–126
rotation testing, 131
safety and, 48–49, 60–61
for short-circuits, 103–104, 127–128, 128
solenoid coil, 190
transformer, 104
trend plot testing, 147–154, 158
voltage testing, 100–102, 115–116, 117, 162
Testing equipment, 55–70, 97. *See also* Circuit testing; *individual meters*
ammeters, 115–120, 127
choosing, 67–71
computer-enhanced DDM, 110–114
computers and, 109
cost of, 114–115
electronic advances in, 54–55, 75
megohmmeter, 123–128
moving-coil volt-ohmmeter, 54, 75
multimeter, 60–66
neon lamps, 136–137
phase meters, 130–131
product information, 68, 70, 71, 73, 107–108, 109–110, 115
programmable logic controller (PLC), 74–75
safety with, 7–9, 57, 60–61, 106
solid-state, 114
verification of, 56–57, 99
wire sorters, 132–133
Thermocouple, 112, 178
Three-phase, 130–131, 149
safety standards for, 9, 130–131, 149
Timed contacts, 23
Timer motor, 28, 104
Timers, 2, 23–24, 94–95
Toggle switch, 16
Trade licensing, 176–177, 180–181, 182
Transformer, 6
diagram of, 35
isolation, 104
primary, 56
safety standards for, 9
secondary, 36, 56
step-down, 105
testing of, 104
Transient, 8
Transient overvoltages, 7–8
Trend plot, 107, 140. *See also* Waveform
connection resistance, 147–150
monitoring mechanical functions, 152–154
motor contact/relay points, 150–152
motor starting current, 147
oscilloscopes, 147–154, 158
testing, 147–154, 158

Trigger circuit, 24. *See also* Initiate circuit
Troubleshooting (conventional), 1–12, 50, 52, 63, 67. *See also* Electrician; On-line troubleshooting; Safety
ammeter, 115–120, 127
with assistant (and radio) 135, 162
control panel, 46–49, 50, 70–73, 88–89
cost effectiveness, 174–175, 181
creative (innovative), 136
definition of, 2–4
documentation of, 80
effective, 174–176, 181
fuses and, 170–171
labeling, 95–96
ladder diagram and, 34, 36–37, 44–45, 70, 71, 80–86, 90
machine malfunction, 84–95, 98, 167
with multimeter, 100–102
neon lamp, 136–137
oscilloscopes, 140, 156–157
performance-over-time monitoring, 140, 156–157, 159
physical equipment, 76–81
plant power quality, 166–167
power check, 80
preventive measures, 81–82
quality assurance and, 156
safety, 45, 48, 52, 86, 97, 98
tools needed for, 84
visual examination, 95–96
without diagrams, 162
Twelve-volt system, 35–36

V

VA. *See* Apparent power
Valve, 187–188. *See also names of valves*
covers, 194
envelopes, 188, 194
for hydraulic systems, 185, 186, 187–188, 191–192, 194
operation of, 188, 194
Valve, spool, 119
Variable frequency drive (VFD), 145–146, 147
Variable speed drive, 7, 129, 140, 155
Viscosity, 192
Volt (V), 5
480-volt installations, 39–40
line conductor diagram, 39
on-line troubleshooting, 9–10
120-volt control circuits, 45
12-volt system, 35–36
240-volt installations, 39–40
Volt-ohmmeter (VOM), 3, 8, 50, 55–56, 58–59, 60, 66, 74, 75, 94
capacitor testing, 105–106, 112
volt sensor, 87–88, 99
Voltage:
control circuit, 39–40, 45
control of, 5, 47, 88

high-voltage installations, 39–40
rating, 8
status, 61, 75
Voltage testing, 100–102, 115–116, 117, 162
 and continuity test, 45, 46, 47, 61–62
 with multimeter, 100–102, 115
 for voltage value, 115–116
Voltage value, 58, 61, 63, 68, 88, 103, 115–116, 136, 152

W

Water pump, 5
Watt (W), 152
 power quality analyzer, 166
Waveform:
 computer displays, 109
 nonsinusoidal, 67
 oscilloscopes, 139–140, 142–147, 158
 technical assistance, 155–156
Wire identification, 132–133, 137, 164–165, 172
Wire magnet, 122
Wire numbers, 11, 25–26, 30, 163. *See also* Line numbers
 testing for continuity, 163–164
Wire sorters, 132–133, 163
 product information, 133, 163
Wiring, 164–168
 add-on, 39
 diagram, 26, 41–43, 97, 172
 and ladder diagram, 32–33, 34
 removing old, 136
Withstand capability, 8